Lecture Notes in Mathematics

Edited by A. Dold and B. Eckmann

1187

Yves Diers

Categories of Boolean Sheaves of Simple Algebras

Springer-Verlag
Berlin Heidelberg New York Tokyo

Author

Yves Diers
Département de Mathématiques
Université de Valenciennes et du Hainaut Cambrésis
59326 Valenciennes – Cedex, France

Mathematics Subject Classification (1980): 18B99, 18F20, 18C10, 18C20, 18D20, 16A90, 16A30, 16A32, 16A74, 20E18, 20F29, 03C90, 03C05

ISBN 3-540-16459-6 Springer-Verlag Berlin Heidelberg New York Tokyo
ISBN 0-387-16459-6 Springer-Verlag New York Heidelberg Berlin Tokyo

Printed in Germany

Printing and binding: Beltz Offsetdruck, Hemsbach/Bergstr.
2146/3140-543210

PREFACE

This book studies and classifies categories of commutative regular algebras, with or without unit, equipped with structure such as order, lattice-order, differential structure, continuous group representation, with integral, algebraic, or separable elements, etc. It is the continuation and development of the study of Locally Finitely Presentable Categories done by P. Gabriel and F. Ulmer, directed mainly towards the study of categories of regular algebras. It uses the axiomatic method that deals with categories equipped with structure instead of categories given concretely by their objects and morphisms. The axiomatic method allows one to give a unified treatment of these categories, to highlight their specific features and to get a fine classification based on universal properties. It gives a construction of the localisation and globalisation processes which leads to a single proof of numerous representation theorems of regular algebras by sheaves. It is shown how well known categorical constructions can be used to get new sectional representations of algebras, algebraic closure constructions, and Galois theories. Categories of continuous representations of finite, or profinite groups, in algebraic categories, and in the category of Boolean algebras in particular, are characterized by universal properties. There is a proof that the notion of commutative regular algebra without unit is 2-equivalent to the notion of commutative unitary regular algebra. All the sheaves of algebras which appear here being based on Boolean or locally Boolean spaces, the passage from stalks to global sections preserves the validity of numerous formulas, and leads to transfer theorems as used by logicians. In order to achieve this work, a lot of new notions have been introduced that will enrich the langage of category theory.

I am indebted to :

Marie-Claude Porter and Timothy Porter, University College of North Wales, who came to my help and translate the manuscript into English,

Madame Claudine Evrard, for the excellent job she has done in typing the manuscript,

The Université de Valenciennes et du Hainaut Cambrésis, for its financial support.

CONTENTS

INTRODUCTION

This work follows in the same direction as the work of A. Grothendieck on Abelian categories and categories of sheaves, of F.W. Lawvere on algebraic categories and categories of sets, of F.E.J. Linton on varietal categories and of P. Gabriel and F. Ulmer on locally presentable categories. It is the development and continuation of the study of locally finitely presentable categories, directed mainly towards the study of categories of regular commutative algebras of all kinds and excluding straight away all categories of modules and Abelian categories. We use the axiomatic method which consists in studying categories with specified structure, rather than categories described concretely by their objects and their morphisms. The axiomatic method allows one to unify the study of these categories, bringing out their specific properties and obtaining a fine classification based on properties of limits and colimits. Each class of categories studied possesses at least one representation theorem, which allows one to describe, in a reasonably concrete way, the categories which belong to it.

Algebraists will see in this work a classification of regular algebras resting solely on some universal properties. The axiomatic presentation allows one to describe the processes of localisation and globalisation and to give a single proof for a multitude of theorems on representations of algebras by sheaves of simple algebras. One can then use general constructions of categories to give new representations of algebras by sheaves. This axiomatic presentation allows one to describe Galois theory and its extensions, using the theory of monads, i.e. of triples. It also allows one to deduce properties of non unitary regular algebras and in particular, their representation by sheaves, from those of regular unitary algebras. We show, in fact, that the two notions of regular unitary algebras and of not necessarily unitary regular algebras are mathematically equivalent.

Logicians will be interested by the fact that all the sheaves of algebras that we use, necessarily have as base a Boolean, or locally Boolean, topological space. One knows that, in this case, the passage from the stalks to global sections, preserves the validity of numerous formulae and gives rise to theorems on the transfer of properties, such as model completion , ω-categoricity, elimination of quantifiers and decidability.

The axiomatic description of categories of continuous representations of a finite or profinite group in any locally presentable category, and in the category of Boolean algebras in particular, may be of interest to group theorists.

The maximal spectrum of a unitary regular commutative ring is a Boolean topological space and its structural sheaf is a sheaf of commutative fields, while the set of continuous global sections of a sheaf of commutative fields on a Boolean topological space is a unitary regular commutative ring. These two correspondences, in fact, set up an equivalence between the category of unitary regular commutative rings and the category of sheaves of commutative fields on Boolean topological spaces, i.e. the category of Boolean sheaves of commutative fields. One obtains an analogous result for each of the following classes of rings : strongly regular rings, commutative regular rings, which are ordered, or lattice-ordered, or differential, or differential and differentially perfect, strongly lattice-ordered projectable rings Baer rings, E-rings, regular K-algebras, all varieties of regular K-algebras, etc.

We resolve the problem of the classification of categories of this type, that is, categories equivalent to categories of sheaves of simple algebraic structures on Boolean topological spaces, i.e. categories of Boolean sheaves of simple algebras, by elaborating an axiomatic description of these categories under the name : locally simple categories. In these categories, one can develop a means of localising each algebra to give a family of simple algebras making up a sheaf on the maximal spectrum, as well as a globalisation process which rebuilds each algebra from its simple localisations. The algebras considered are mostly unitary algebras over a unitary ring, with supplementary algebraic structures such as being regular algebras, Baer algebras, lattice-ordered algebras, regular differential algebra, etc. However, the results can actually be applied to a much bigger class of algebras comprising universal algebras, algebras in the sense of Lawvere or Bénabou, algebras for a monad of finite rank in the sense of Eilenberg-Moore, or algebras for a theory in the sense of P. Gabriel and F. Ulmer.

Morphisms of locally simple categories allow one to link the locally simple categories amongst themselves, to construct locally simple categories by the general methods for constructing categories and to characterise these constructions. One thus constructs products of locally simple categories, full locally simple subcategories, locally simple categories of fractions, locally simple categories of objects over an object, locally simple categories of algebras for an algebraic theory or for a monad, locally simple categories of coalgebras for a comonad, of representations of a finite or profinite group, and locally simple categories of functors. One obtains in this way theorems on representations of algebras by sheaves of simple algebras.

Locally Noetherian locally simple categories, locally preGalois or locally Galois categories, form particular classes of locally simple categories, containing the categories of algebras which are algebraic over a field, and categories of separable algebras. In these, one can construct algebraic closures, profinite fundamental groups, fundamental functors and one has a Galois theory containing all the classical Galois theories and their extensions, by using the theory of monads.

For non unitary commutative regular rings, one must replace Boolean spaces by locally Boolean topological spaces and the set of continuous global sections by the set of continuous global sections with compact support. The correspondences establish that the category of not necessarily unitary commutative regular rings is equivalent to the category of sheaves of commutative fields on locally Boolean topological spaces, i.e. to the category of locally Boolean sheaves of commutative fields.

We also resolve the problem of classifying categories of this type, that is to say, categories equivalent to categories of sheaves of simple algebraic structures on locally Boolean spaces, i.e. categories of locally Boolean sheaves of simple algebras, by working out an axiomatic description of these categories under the name : quasi-locally simple categories. We bring to the fore a solution, at least for regular algebras, to the conflict opposing partisans of unitary algebras for whom the study of non-unitary algebras is without interest and even without object, since it is always possible to add a unity element formally, to supporters of not necessarily unitary algebras, who maintain that the addition of a unity element can be difficult, that it is useless and that algebras of continuous functions <u>of</u> compact support, as well as ideals, are not necessarily unitary algebras, which are worthy of attention. Our solution consists in proving that the notion of not necessarily unitary regular algebra is mathematically equivalent to the notion of unitary regular algebra.

The two notions of unitary and not necessarily unitary regular algebra are not equivalent in the usual sense. If one may, on the one hand, forget the unity element of a unitary algebra and on the other, add a unit formally to a non unitary algebra, the correspondences thus obtained are not mutually inverse to each other, not even up to isomorphism. Moreover, between two unitary algebras, the non-unitary homomorphisms are noticeable different from unitary homomorphisms. In other words, a category of not necessarily unitary regular algebras is not equivalent to a category of unitary regular algebras. One can easily prove this by noting that categories of the first type have a null object whilst those of the second have a strict final object. We maintain, however, that the notions of unitary regular algebra and not necessarily unitary regular algebras are 2-equivalent, that is to say that the notions of category of unitary regular algebras and category of not necessarily unitary regular algebra are equivalent, in other words, that there exists a one-to-one correspondence up to equivalence of categories, between the two types of categories. More precisely, we prove that the 2-category of categories of unitary regular algebras is 2-equivalent to the 2-category of categories of not necessarily unitary regular algebras. We use this result to deduce properties of varieties of not necessarily unitary regular algebras and, in particular, the representation by sheaves of simple algebras on locally Boolean spaces, from those of varieties of unitary regular algebras.

<u>In Chapter 1</u> : <u>Categories of Boolean sheaves of algebras</u>, we recall the axiomatic description of categories of sets with an algebraic structure, given by P. Gabriel and F. Ulmer under the name : Locally Finitely Presentable Categories, [16], and we study, axiomatically, categories of Boolean sheaves of algebraic structures under the name : Locally Indecomposable Categories [15].

A locally indecomposable category is a locally finitely presentable category in which finite products are codisjoint and couniversal and the product of the initial object, Z, by itself is finitely presentable. In such a category $\mathbb{A}$, for each object A, the quotient objects of A of the form $\delta : A \to B$ such that there is a morphism $\delta' : A \to B'$ making up with δ, a product $(\delta : A \to B, \delta' : A \to B')$, are called the direct factors of A. Ordered by the order of quotient objects of A, they constitute a Boolean algebra, denoted $\Delta(A)$, in which the two quotient objects δ and δ' are complementary. The direct factors are coclassified by the object $Z^2 = Z\times Z$, which is finitely presentable, in other words, the object Z^2 is the coclassifier for direct factors, that is to say that if $\pi : Z^2 \to Z$ is the first projection and $i_B : Z \to B$ is the unique morphism, then for each direct factor $\delta : A \to B$, there is a unique morphism $f : Z^2 \to A$ such that (i_B,δ) is the amalgamated sum of (π,f). For example, in the category $\mathbb{R}$ngc of unitary commutative rings, the square, $\mathbb{Z}^2$, of the initial object $\mathbb{Z}$ is isomorphic to the quotient ring $\mathbb{Z}[X]/(X^2-X)$ and the Boolean algebra, $\Delta(A)$, is isomorphic to the Boolean algebra of idempotents of A.

The indecomposable spectrum of an object A in $\mathbb{A}$ is the Stone space of the Boolean algebra $\Delta(A)$. It is denoted $\mathrm{Spec}_{\mathrm{Ind}}(A)$. The object A can be localised, at each point of its indecomposable spectrum, to give a quotient object which is indecomposable as a product of two objects. The set of these localisations of A coincides with the set of maximal indecomposable quotient objects of A, called the indecomposable components of A. It is, in fact, the set of stalks of a sheaf, $\widetilde{A}$, defined on $\mathrm{Spec}_{\mathrm{Ind}}(A)$ with values in $\mathbb{A}$, whose object of global sections is A. We call $\widetilde{A}$, the structural sheaf on $\mathrm{Spec}_{\mathrm{Ind}}(A)$.

The correspondence which, to an object A, associates the pair $(\mathrm{Spec}_{\mathrm{Ind}}(A),\widetilde{A})$ defines, in fact, a functor $\Sigma : \mathbb{A} \to \mathbb{S}\mathrm{hvBoolInd}\,\mathbb{A}$ from the category $\mathbb{A}$ to the category $\mathbb{S}\mathrm{hvBoolInd}\,\mathbb{A}$ of sheaves of indecomposable objects of $\mathbb{A}$ on Boolean topological spaces. The functor Σ is an equivalence of categories for which the quasi-inverse functor is the global section functor. Thus one obtains a theorem on the representation of locally indecomposable categories by categories of sheaves of algebraic structures on Boolean topological spaces.

A morphism of locally indecomposable categories is a functor $U : \mathbb{A} \to \mathbb{B}$, whose source, $\mathbb{A}$, and target, $\mathbb{B}$, are two locally indecomposable categories, which preserves filtered colimits and which has a left adjoint preserving finite products. Such a morphism induces a bijection between the direct factors of an object A of $\mathbb{A}$ and those of UA. The Boolean algebras $\Delta(A)$ define, in fact, a morphism of locally indecomposable categories $\Delta : \mathbb{A} \to \mathbb{B}\mathrm{ool}$ and the category $\mathbb{B}\mathrm{ool}$ turns out to be the 2-final locally indecomposable category. Full locally indecomposable subcategories of a locally indecomposable category $\mathbb{A}$ are those full subcategories of $\mathbb{A}$ for which the inclusion functor is a morphism of locally indecomposable categories. These are also the full reflexive subcategories $\mathbb{B}$ of $\mathbb{A}$ closed under filtered colimits and satisfying one of the following conditions :

(1) $\mathbb{B}$ is closed under direct factors

(2) an object of $\mathbb{A}$ is in $\mathbb{B}$ if, and only if, its indecomposable components are in $\mathbb{B}$. Every variety of a locally indecomposable category of algebras generates a full locally indecomposable subcategory. If Σ is a set of finitely presentable morphisms in a locally indecomposable category $\mathbb{A}$, one can construct the locally indecomposable category of fractions of $\mathbb{A}$ defined by Σ. In the case where the morphisms of Σ have, as sources, weakly initial objects, this category of fractions is equivalent to the full locally indecomposable subcategory of $\mathbb{A}$ having as objects, the objects, A, which are left closed for Σ, i.e. such that the functions $\mathrm{Hom}_{\mathbb{A}}(\sigma,A)$ are bijective for all $\sigma \in \Sigma$. In the general case, this category of fractions is equivalent to the full locally indecomposable subcategory of $\mathbb{A}$ having as objects, the objects, A, which are locally left closed for Σ, i.e. for which all the indecomposable components are left closed for Σ. If Γ is a set of finite families of finitely presentable morphisms having the same source in a locally indecomposable category $\mathbb{A}$, one can construct the universal locally indecomposable category in which the families of Γ become products. In the case where the morphisms in the families of Γ have, as sources, weakly initial objects, this category is equivalent to the full subcategory of $\mathbb{A}$ whose objects are strictly left closed for Γ, i.e. they make the functions $\langle \mathrm{Hom}_{\mathbb{A}}(Y_i,A)\rangle : \coprod_{i=1}^{n} \mathrm{Hom}_{\mathbb{A}}(X_i,A) \to \mathrm{Hom}_{\mathbb{A}}(X,A)$ into bijections for every family $(Y_i : X \to X_i)_{i\in[1,n]}$ of Γ. In the general case, this category is equivalent to the full subcategory of $\mathbb{A}$ of which the objects are locally strictly left closed for Γ, i.e. for which the indecomposable components are strictly left closed for Γ. If $\mathbb{T}$ is a monad of finite rank, which preserves finite products, on a locally indecomposable category $\mathbb{A}$, the category $\mathbb{A}^{\mathbb{T}}$ of $\mathbb{T}$-algebras is locally indecomposable and the functor forgetting structure, $U^{\mathbb{T}} : \mathbb{A}^{\mathbb{T}} \to \mathbb{A}$ is a morphism of locally indecomposable categories. If $\mathbb{T}$ and $\mathbb{T}'$ are two algebraic theories in the sense of F.W. Lawvere and $U : \mathbb{A}\mathrm{lg}(\mathbb{T}') \to \mathbb{A}\mathrm{lg}(\mathbb{T})$ is an algebraic functor between the corresponding categories of algebras, then U is a morphism of locally indecomposable categories if, and only if, the category $\mathbb{A}\mathrm{lg}(\mathbb{T})$ is locally indecomposable and U lifts direct factors. If $\mathbb{A}$ is a locally indecomposable category and A is an object of $\mathbb{A}$, the category $A/\mathbb{A}$ of objects of $\mathbb{A}$ under A, is locally indecomposable.

In chapter 2 : Categories of Boolean Sheaves of Simple Algebras, we study locally simple categories ; this will be our name for categories of Boolean sheaves of algebraic structures with simple stalks, i.e. having no non trivial regular quotient.

A locally simple category is defined to be a locally finitely presentable category in which finite products are codisjoint and couniversal, the product , Z^2, of the initial object Z with itself is finitely presentable, and the functor $Hom(Z^2,-)$ reflects monomorphisms. It is a locally indecomposable category. In such a category the finitely presentable objects are codecidable, i.e. their codiagonal is a complemented quotient object. In fact, all the finitely generated regular quotient objects are complemented and are direct factors. It follows that the objects which are indecomposable as a product of two objects are exactly the simple objects.

If $\mathbb{C}$ is a finitely complete small category with universal disjoint finite sums and in which objects are decidable, the category, $\mathbb{C}ont_{\aleph_0}[\mathbb{C},\mathbb{E}ns]$, of finitely continuous functors from $\mathbb{C}$ to $\mathbb{E}ns$, the category of sets, is locally simple, and its simple objects are those finitely continuous functors which preserve finite sums. This can be applied, for example, to the case where $\mathbb{C}$ is a small Boolean topos such as the topos of finite cardinals of a topos [23,6.2.] or a small atomic topos [3]. One obtains a first representation of locally simple categories by proving that any locally simple category is equivalent to a category of this form, $\mathbb{C}ont_{\aleph_0}[\mathbb{C},\mathbb{E}ns]$.

A locally simple category is algebraic in the sense of Gabriel-Ulmer [16], and thus is equivalent to an I-algebraic category in the sense of Bénabou [6], i.e. to a category of families of sets with an algebraic structure defined by internal or external operations and some identities between composed operations. One thus obtains a second representation of locally simple categories. A locally simple category is, furthermore, varietal in the sense of Linton [31], and thus is monadic [33] over the category of sets. A locally simple category is algebraic in the sense of Lawvere [30] if, and only if, it has a finite proper generating set made of finitely presentable objects. In this case, we will say it is locally simple algebraic. These results may seem surprising, considering some examples of locally simple categories. They have lead us to make precise a simple algebraic description of commutative Baer rings and of projectable strongly lattice-ordered commutative rings. In the general case, there are not always privileged generating objects and consequently, the concept of "underlying set" is not canonically defined.

A locally simple category $\mathbb{A}$ is regular and exact in the sense of Barr [2]. For each object A of $\mathbb{A}$, the equivalence relations on A form a lattice E(A) isomorphic to the dual of the lattice of regular quotient objects of A. This lattice is distributive, complete, compactly generated and its compact elements are the finitely generated equivalence relations and these coincide with the complemented elements. It is a complete Heyting algebra. The set of proper maximal equivalence relations on A gives the maximal spectrum of A, denoted $\mathrm{Spec}_{Max}(A)$. With the Stone topology, it is a Boolean topological space whose lattice of open sets is isomorphic to E(A). The object A can be localised at each point, R, of its maximal spectrum, giving the simple quotient object A/R. The set of these localised objects is the set of stalks of a sheaf $\tilde{A}$ with base $\mathrm{Spec}_{Max}(A)$, with values in $\mathbb{A}$. We call it the structural sheaf on $\mathrm{Spec}_{Max}(A)$. The object A is then isomorphic to the object of global sections of $\tilde{A}$.

By associating to each object A of $\mathbb{A}$, the pair $(\mathrm{Spec}_{Max}(A),\tilde{A})$, one defines a functor $\Sigma : \mathbb{A} \to \mathbb{S}\mathrm{hvBoolSim}\,\mathbb{A}$ from $\mathbb{A}$ to the category of sheaves of simple objects of $\mathbb{A}$ on Boolean topological spaces. This functor is an equivalence of categories whose quasi-inverse functor is the global section functor. One thus obtains a third representation of locally simple categories : by categories of sheaves of simple algebras on Boolean spaces.

A morphism of locally simple categories is a functor $U : \mathbb{A} \to \mathbb{B}$ whose source and target are two locally simple categories, which preserves filtered colimits and which has a left adjoint which preserves finite products. It is a morphism of locally indecomposable categories which is exact, preserves and reflects simple objects, reflects monomorphisms and the final object, and which lifts uniquely equivalence relations and regular quotient objects. Let $\mathbb{A}$ be a locally simple category. If A is an object of $\mathbb{A}$, the target functor $A/\mathbb{A} \to \mathbb{A}$ is a morphism of locally simple categories. If $\mathbb{T}$ is a monad of finite rank preserving finite products on $\mathbb{A}$, the functor $U^{\mathbb{T}} : \mathbb{A}^{\mathbb{T}} \to \mathbb{A}$ is a morphism of locally simple categories. The monadic morphisms of locally simple categories are characterised by specific "Beck theorems" : a morphism of locally simple categories is monadic if, and only if, it reflects isomorphisms. A functor $U : \mathbb{B} \to \mathbb{A}$ is a monadic morphism of locally simple categories if, and only if, the category $\mathbb{B}$ has generalised kernel pairs and the functor U (i) has a left adjoint, (ii) preserves filtered colimits and (iii) creates regular quotient objects. A locally simple subcategory of $\mathbb{A}$ is defined as being a subcategory of $\mathbb{A}$, the inclusion functor of which is a morphism of locally simple categories. Such a subcategory is closed under regular quotients and an object of $\mathbb{A}$ belongs to it if, and only if, its simple regular quotients belong to it. Locally simple categories of fractions of $\mathbb{A}$ appear as locally simple full subcategories of $\mathbb{A}$. Varieties in $\mathbb{A}$ are particular locally simple full subcategories of $\mathbb{A}$. This notion of variety is naturally defined since the category $\mathbb{A}$ is equivalent to an I-algebraic category. One shows that the varieties in $\mathbb{A}$ are exactly the full subcategories of $\mathbb{A}$ which are closed under products, subobjects and regular quotients, that they are generated by their simple objects and that they are in one-to-one correspondence with the full subcategories of the category, $\mathbb{S}\mathrm{im}\,\mathbb{A}$, of simple objects of $\mathbb{A}$, closed under ultraproducts and subobjects. If Γ is a set of finite families of finitely presentable regular epimorphisms having the same source in $\mathbb{A}$, one can construct the universal locally simple category in which the families in Γ become counions of quotient objects, and this category is a variety in $\mathbb{A}$.

A locally simple algebraic theory is an algebraic theory in the sense of Lawvere having at least two constants denoted 0 and 1 and a quaternary comparison operation, C, in the sense of Kennison [26], which satisfies the axiom : $C(0,1,X,0) = 0$. The category of algebras for such a theory is locally simple. One obtains a fourth representation of locally simple categories by showing that each locally simple algebraic category is equivalent to a category of algebras for such a theory.

If $\mathbb{T}$ is an algebraic theory in the sense of Lawvere having two distinguished constants 0 and 1, a sheaf of $\mathbb{T}$-algebras on a topological space X is said to be totally separated if the space X and the étale space associated to the sheaf are separated and $0 \neq 1$ in each stalk. The category, $\mathbb{S}\mathrm{hvBoolTsep}\mathbb{A}\mathrm{lg}(\mathbb{T})$, of totally separated Boolean sheaves of $\mathbb{T}$-algebras is locally simple algebraic and is equivalent to the category of $\mathbb{CT}$-algebras where $\mathbb{CT}$ denotes the locally simple algebraic theory obtained by adjoining to $\mathbb{T}$ a comparison operation. The simple objects can be identified as being $\mathbb{T}$-algebras. If $\mathbb{M}$ is a full subcategory of $\mathbb{A}\mathrm{lg}(\mathbb{T})$ closed under ultraproducts and subobjects, the category, $\mathbb{S}hvBoolTsep\,\mathbb{M}$, of totally separated Boolean sheaves of algebras in $\mathbb{M}$ is also locally simple algebraic and the categories of this form are precisely the varieties in $\mathbb{S}hvBoolTsep\mathbb{A}lg(\mathbb{T})$. One obtains a fifth representation of locally simple categories on proving that the locally simple algebraic categories are precisely the categories equivalent to categories of the form $\mathbb{S}hvBoolTsep\,\mathbb{M}$. These later results link with the work of Kennison on triples and representations by compact sheaves [26].

Categories of regular algebras algebraic over a field, K, are locally simple categories which are comonadic over categories of regular K-algebras. However, categories of coalgebras for a comonad on a locally presentable category are not, in general, locally presentable. One thus is forced to make precise the notion of comonad being used. A comonad $\mathbb{L} = (L,\varepsilon,\delta)$ on a locally finitely presentable category $\mathbb{A}$ is said to be of finite rank if the functor L preserves filtered colimits and if each morphism, $f : X \to A$, from a finitely presentable object X of $\mathbb{A}$ to an object, A, underlying an $\mathbb{L}$-coalgebra, (A,a), factorises in the form $f = hg$, where $g : X \to B$ is a morphism whose target B is finitely presentable and h is a morphism underlying an $\mathbb{L}$-coalgebras morphism $h : (B,b) \to (A,a)$. Under these conditions, the category $\mathbb{A}^{\mathbb{L}}$ of $\mathbb{L}$-coalgebras is locally finitely presentable and its finitely presentable objects are precisely the $\mathbb{L}$-coalgebras on finitely presentable objects of $\mathbb{A}$. If, in addition, one assumes that the category $\mathbb{A}$ is locally simple and that $\mathbb{L}$ preserves finite products, then the category $\mathbb{A}^{\mathbb{L}}$ is locally simple. The category, $\mathbb{A}lgcAlgReg(k)$, of regular algebraic commutative k-algebras as well as the category $\mathbb{A}lgcAlgSep(k)$ of separable algebraic commutative k-algebras are comonadic for an idempotent comonad of finite rank, which preserves finite products, on the category $\mathbb{A}lgcReg(k)$ of regular commutative k-algebras.

Categories of continuous representations of a profinite group, G, are also categories of coalgebras for a comonad of finite rank. To start with we should point out that, for any locally finitely presentable category $\mathbb{A}$, the sets $Hom_{\mathbb{A}}(A,B)$ canonically have a topology, in such a way that $\mathbb{A}$ becomes a category enriched over the category of zero dimensional topological spaces. It follows that the automorphism groups, Aut(A) of objects A of $\mathbb{A}$ are topological groups having a basis of open neighbourhoods of the identity element, made up of the Galois subgroups $G_u = \{\sigma \in Aut(A) : \sigma u = u\}$ for the various morphisms $u : X \to A$ with finitely presentable source and with target A. If G is a topological group, a continuous representation of G on $\mathbb{A}$ is a pair (A,α) consisting of an object A of $\mathbb{A}$ and a continuous map, $\alpha : G \to \mathrm{Aut}_{\mathbb{A}}(A)$. These representations form, with their morphisms the category $\mathbb{A}^{\mathrm{Cont}(G)}$. If G is a profinite group, this category is comonadic for a comonad of finite rank on $\mathbb{A}$. It is thus locally finitely presentable. It is locally simple if $\mathbb{A}$ is. For example, the category $\mathbb{B}\mathrm{ool}^{\mathrm{Cont}(G)}$ of continuous representations of a profinite group G in Boolean algebras is a locally simple category whose dual is equivalent to the category G-$\mathbb{E}$spBool of Boolean G-spaces.

Locally Noetherian locally simple categories have a special structure. The Noetherian objects of a category are defined by Gabriel and Ulmer, [16] as being the objects, A, such that every decreasing sequence of extremal quotient objects of A is stationary, and a category $\mathbb{A}$ is said to be locally Noetherian if it is cocomplete and has a proper generating set made up of Noetherian finitely presentable objects whose finite sums are still Noetherian, or, equivalently, if it is a locally finitely presentable category in which the finitely generated objects are finitely presentable. A locally simple category is locally Noetherian if, and only if, its finitely presentable objects are semi-simple. The categories, $\mathbb{A}$lgcAlgReg(k), $\mathbb{A}$lgAlgSep(k), and $\mathbb{B}$ool are locally Noetherian, locally simple categories. The category of coalgebras for a comonad of finite rank on a locally Noetherian category is locally Noetherian. The category of continuous representations of a profinite group in a locally Noetherian category is locally Noetherian. The following construction of locally Noetherian locally simple categories is generic. Let $\mathbb{K}$ be a small monomorphic category with well finite multisums, i.e. a small category in which all the morphisms are monomorphisms and such that, for any finite family $(X_i)_{i\in I}$ of objects, there exists a finite family, $(\gamma_{ji} : X_i \to Y_j)_{(i,j)\in I\times J}$, of morphisms such that each family $(f_i : X_i \to Y)_{i\in I}$ of morphisms with the same target Y, factorises uniquely through a single family $(\gamma_{ji})_{i\in I}$. Then the category $\mathbb{F}$amfin $\mathbb{K}$ of finite families of objects in $\mathbb{K}$ is finitely cocomplete, has codisjoint and couniversal finite products and its objects are codecidable. One proves that the category, $\mathbb{C}\mathrm{ont}_{\aleph_0}[\mathbb{F}\mathrm{amfin}\,\mathbb{K}^{op}, \mathbb{E}\mathrm{ns}]$, of finitely continuous contravariant functors defined on $\mathbb{F}$amfin $\mathbb{K}$ and with values in $\mathbb{E}$ns, is a locally Noetherian locally simple category whose category of finitely presentable simple objects can be identified to $\mathbb{K}$. For example, if $\mathbb{K}$ is the category of algebraic extension fields of finite degree of a commutative field, k, the category $\mathbb{C}ont_{\aleph_0}[\mathbb{F}amfin\ \mathbb{K}^{op}, \mathbb{E}ns]$ is equivalent to the category $\mathbb{A}$lgReg(k). It is remarkable that any locally Noetherian locally simple category $\mathbb{A}$ can be entirely rebuilt starting from its finitely presentable simple objects and that it is equivalent to the category, $\mathbb{C}ont_{\aleph_0}[\mathbb{F}amfin\ \mathbb{K}^{op}, \mathbb{E}ns]$ where $\mathbb{K}$ is the category of finitely presentable simple objects of $\mathbb{A}$. One thus obtains a sixth representation of locally simple categories.

A category $\mathbb{A}$ satisfies the amalgamation property if, for any pair of monomorphisms with the same source $(f : A \to B,\ g : A \to C)$, there exists a pair of monomorphisms $(m : B \to D,\ n : C \to D)$ satisfying $mf = ng$. A locally simple category $\mathbb{A}$ satisfies the amalgamation property if, and only if, the category, $\mathbb{S}im\ \mathbb{A}$, of simple objects of $\mathbb{A}$ satisfies it, and this is equivalent to the fact that monomorphisms are couniversal in $\mathbb{A}$. Thus, for example, the amalgamation property for fields implies that monomorphisms are couniversal in the categories $\mathbb{R}$ngcReg, $\mathbb{A}$lgcReg(k), $\mathbb{A}$lgcAlgReg(k), and $\mathbb{A}$lgcAlgSep(k).

The objects of finite power in a locally presentable category $\mathbb{A}$, are defined as being those objects, A, such that the sets $Hom_{\mathbb{A}}(B, A)$ are finite for all finitely presentable objects B of $\mathbb{A}$. In categories algebraic on a base set, these are the finite algebras, but in categories algebraic on an infinite family of base sets, this is not the case. In the category $\mathbb{A}$lgcAlgReg(k), the algebraic extension fields of k are simple objects of finite power and this property expresses the fact that a non zero polynomial can only have a finite number of zeros in a field. For each object, A, of finite power of $\mathbb{A}$, the topological spaces $Hom_{\mathbb{A}}(B,A)$ are Boolean and define a functor $Hom_{\mathbb{A}}(-,A) : \mathbb{A}^{op} \to \mathbb{E}spBool$, which has a left adjoint, called the Boolean exponential functor with base A, whose value for a Boolean space X is the Boolean power, A^X, of A by X. Moreover, the automorphism group of an object of finite power is a profinite group.

Locally preGalois categories are the locally simple categories in which the initial object is simple, monomorphisms are couniversal and simple objects are of finite power. For example the category $\mathbb{A}$lgcAlgReg(k) is locally preGalois. These categories are locally Noetherian categories in which the objects of finite power are precisely the semi-simple objects. One proves that a locally preGalois category, $\mathbb{A}$, has a maximum simple object, L, i.e. a simple object L such that every simple object of the category is a subobject of L. It is precisely the algebraic closure of k in the category $\mathbb{A}$lgcAlgReg(k). The profinite group, Aut(L), of automorphisms of L, is the fundamental group, $G_{\mathbb{A}}$, of $\mathbb{A}$, and the functor $Hom_{\mathbb{A}}(-,L) : \mathbb{A}^{op} \to$ $\mathbb{E}$sp$\mathbb{B}$ool is the fundamental functor of $\mathbb{A}$. This functor determines an adjunction $\mathbb{A}^{op} \rightleftarrows \mathbb{E}$spBool, which generates, on the category $\mathbb{E}$sp$\mathbb{B}$ool, the monad canonically associated to the group, $G_{\mathbb{A}}$, and a comparison functor, $H_{\mathbb{A}} : \mathbb{A}^{op} \to$ $G_{\mathbb{A}}$-$\mathbb{E}$sp$\mathbb{B}$ool, which determines an adjunction $\mathbb{A}^{op} \rightleftarrows G_{\mathbb{A}}$-$\mathbb{E}sp\mathbb{B}$ool, which induces an adjunction $(\$im\ \mathbb{A})^{op} \rightleftarrows G_{\mathbb{A}}$-$\mathbb{E}sp\mathbb{B}$ool Homo and a Galois correspondence $(Sim\ \mathbb{A})^{op} \rightleftarrows SF(G_{\mathbb{A}})$ between the set, Sim$\mathbb{A}$, of simple objects of $\mathbb{A}$ and the set $SF(G_{\mathbb{A}})$ of closed subgroups of $G_{\mathbb{A}}$. For instance, for $\mathbb{A}$ = $\mathbb{A}$lgcAlgReg(k), Sim $\mathbb{A}$ is the set, ExtAlg(k), of algebraic extension fields of k, $G_{\mathbb{A}}$ is the Galois group, $G_{\bar{k}}$, of the algebraic closure $\bar{k}$ of k, and the non-bijective Galois correspondence $ExtAlg(k)^{op} \rightleftarrows SF(G_{\bar{k}})$ is classical. If $\mathbb{A}$ is a locally preGalois category whose maximum simple object is L and if N is a normal simple object of $\mathbb{A}$, i.e. such that the group Aut(N) operates transitively on $Hom_{\mathbb{A}}(N,L)$, then the full subcategory, A//N, of $\mathbb{A}$, of objects of $\mathbb{A}$ which are locally over N, i.e. whose simple components are subobjects of N, is a locally preGalois category. For example, if K is a normal algebraic extension field of K, the category $\mathbb{A}$lgcAlgReg(k)//K is the cateogry $\mathbb{A}$lgcAlgReg(k/K) of regular algebraic commutative k-algebras whose localisations are algebraic extension fields intermediate between k and K.

Locally Galois categories are balanced preGalois categories in which equalisers are couniversal. The categories $\mathbb{B}$ool and $\mathbb{A}$lgcAlgSep(k) are locally Galois. If $\mathbb{A}$ is a locally Galois category, the category, $\mathbb{A}^{Cont(G)}$, of continuous representations of a profinite group G in $\mathbb{A}$, and also the category $\mathbb{A}$//N of objects in $\mathbb{A}$ locally over a normal simple object N, are locally Galois. One proves that the fundamental functor, $Hom_{\mathbb{A}}(-,L) : \mathbb{A}^{op} \to$ $\mathbb{E}$sp$\mathbb{B}$ool, of a locally Galois category, $\mathbb{A}$, is monadic and consequently, that the comparison functor $H_{\mathbb{A}} : \mathbb{A}^{op} \to$ $G_{\mathbb{A}}$-$\mathbb{E}$sp$\mathbb{B}$ool is an equivalence of categories. Thus we obtain dualities between :

(0) the category $\mathbb{A}$ and the category $G_{\mathbb{A}}$-$\mathbb{E}$sp$\mathbb{B}$ool,

(1) the category $\mathbb{A}_o$ of finitely presentable objects of $\mathbb{A}$ and the category $G_{\mathbb{A}}$-$\mathbb{E}$nsfin of finite continuous $G_{\mathbb{A}}$-sets,

(2) the category, \$im $\mathbb{A}$, of simple objects of $\mathbb{A}$ and the category, $G_{\mathbb{A}}$-$\mathbb{E}$sp$\mathbb{B}$ool$\mathbb{H}$omo, of homogeneous Boolean $G_{\mathbb{A}}$-spaces,

(3) the ordered set, Sim $\mathbb{A}$, of simple objects of $\mathbb{A}$ and the ordered set $SF(G_{/\mathbb{A}})$ of closed subgroups of $G_{/\mathbb{A}}$,

(4) the category, $\mathbb{S}$im $\mathbb{A}_0$ of finitely presentable simple objects of $\mathbb{A}$ and the category $G_{/\mathbb{A}}$-$\mathbb{E}$nsfin$\mathbb{H}$omo, of homogeneous continuous finite $G_{/\mathbb{A}}$-sets,

(5) the ordered set Sim $\mathbb{A}_0$ of finitely presentable simple objects of $\mathbb{A}$ and the ordered set $SO(G_{/\mathbb{A}})$ of open subgroups of $G_{/\mathbb{A}}$. Each of these dualities is a more or less classical "Galois theory". Duality (5) is the classical Galois theory and duality (3) is the extension of it to extensions of infinite degree. Duality (4) is the abstract "Galois theory" described by M. Barr [4] and duality (2) is its extension to extensions of infinite degree. Duality (1) is the axiomatic "Galois theory" described by A. Grothendieck [20] and duality (0) is its extension to extensions of infinite degree. Finally one proves that locally Galois categories are precisely the categories equivalent to categories of the form, $\mathbb{B}\mathrm{ool}^{\mathrm{Cont}(G)}$, of continuous representations of a profinite group G in Boolean algebras.

<u>In chapter 3</u> : <u>Categories of locally Boolean sheaves of simple algebras</u>, we give under the name : quasi locally simple categories, an axiomatic description of categories of sheaves of algebraic structures having for base locally Boolean topological spaces, i.e. spaces in which each point has a Boolean neighbourhood, and having simple algebras as stalks.

A quasi locally simple category is defined as being a locally finitely presentable category with a null object, in which products are codisjoint, counions of pairs of regular quotient objects are couniversal, and finitely generated regular quotient objects are complemented and coclassified by an object Z. For example, categories of not necessarily commutative rings which are regular, regular and strongly lattice ordered, regular differential, regular lattice ordered and projectable, Baer or E-rings, as well as all categories of associated algebras and their varieties, are quasi locally simple. If $\mathbb{C}$ is a finitely complete small category having disjoint universal finite sums and whose objects are decidable, and if one denotes by $\mathbb{P}$art($\mathbb{C}$), the category of partial morphisms of $\mathbb{C}$, in which objects are those of $\mathbb{C}$ and the morphisms from X to Y are pairs (X,f) made up of a regular subobject X' of X and a morphism $f : X' \to Y$, then the category

$\mathbb{C}\mathrm{ont}_{\aleph_0}[\mathbb{P}\mathrm{art}(\mathbb{C}), \mathbb{E}\mathrm{ns}]$, of finitely continuous functors defined on $\mathbb{P}\mathrm{art}(\mathbb{C})$ with values in $\mathbb{E}\mathrm{ns}$, is quasi locally simple. And one proves that any quasi locally simple category is equivalent to a category of this form. This applies, for instance, in the case where $\mathbb{C}$ is a small Boolean or atomic topos [3]. If $\mathbb{A}$ is a locally simple category, the category $\mathbb{A}^*$ of augmented objects of $\mathbb{A}$, i.e. of objects over the initial object, is quasi locally simple, as is the category $\mathbb{S}$hvLocBool$\mathbb{S}$im $\mathbb{A}$, of locally Boolean sheaves of simple objects of $\mathbb{A}$.

Let $\mathbb{A}$ be a quasi locally simple category. One must distinguish in $\mathbb{A}$, between kernel and equaliser and between cokernel and coequaliser. A morphism is a monomorphism if, and only if, its kernel is null, and it is a regular epimorphism, if, and only if, it is a cokernel. Every finitely generated regular epimorphism is a direct factor, that is to say, is a canonical projection of a product. A monomorphism, or a subobject, is said to be normal if it is the kernel of a morphism. In categories of K-algebras, the normal subobjects are the ideals. By associating to each normal subobject of an object A, its cokernel, and to each regular quotient object of A, its kernel, one establishes a one-to-one correspondence between the normal subobjects of A and its regular quotient objects. The set of normal subobjects of A is a complete compactly generated Heyting algebra in which the compact elements are complemented, are of the form $(1,0) : X \to X \times Y = A$, and give a not necessarily unitary Boolean algebra. A semi-epimorphism is a morphism whose cokernel is null. It defines a semi-quotient object. Any morphism in $\mathbb{A}$ can be factorised in a natural and essentially unique way, into a semi-epimorphism and a normal monomorphism ; it is the normal factorisation.

An object A is said to be of finite type if the functor $\mathrm{Hom}_{\mathbb{A}}(A,-)$ preserves normal monomorphic filtered colimits, i.e. those in which the canonical injections are normal monomorphisms. These objects are precisely the semi-quotients of the finitely presentable objects. The coclassifying object, Z, is the generic object of finite type in the sense that any object of finite type is a semi-quotient of Z in a unique way. In categories of regular K-algebras, the objects of finite type are exactly the unitary algebras. In categories of locally Boolean sheaves of simple algebras, they are exactly the Boolean sheaves. The objects of finite type in $\mathbb{A}$, together with the semi-epimorphisms make up the category $\mathbb{A}_{tf}$. It is a reflexive subcategory of $\mathbb{A}$, which is closed under connected colimits. One proves that it is locally simple and that the category $\mathbb{A}$ is equivalent to the category $\mathbb{A}^*_{tf}$ of augmented objects in $\mathbb{A}_{tf}$. As a result, quasi locally simple categories are precisely the categories equivalent to categories of augmented objects in locally simple categories. On the other hand, one proves that each locally simple category $\mathbb{B}$ is equivalent to the category of objects of finite type in the category $\mathbb{B}^*$ of augmented objects in $\mathbb{B}$. One thus deduces that locally simple categories are precisely those categories equivalent to categories of objects of finite type in quasi locally simple categories. So as to be able to formulate precisely the links that exist between quasi locally simple categories and locally simple categories, it is necessary to introduce morphisms between these categories.

A morphism of quasi locally simple categories is a functor whose source and target are both quasi locally simple categories, which reflects the null object, preserves filtered colimits and has a left adjoint which preserves finite products. Such a morphism lifts, uniquely, regular quotient objects and normal subobjects. Its adjunction morphisms are semi-epimorphisms and it induces a morphism of locally simple categories between the categories of objects of finite type. A 2-morphism of quasi locally simple categories is a natural transformation $\alpha : U \to V$ whose source and target are two morphisms of quasi-locally simple categories and whose value on the coclassifier, Z, is a semi-epimorphism. Such a 2-morphism induces a 2-morphism of locally simple categories between the categories of objects of finite type. One thus obtains the 2-category, $\mathbb{Q}uaLocSim$, of quasi locally simple categories. One proves that the 2-category, $\mathbb{Q}uaLocSim$, is 2-equivalent to the 2-category, $\mathbb{L}ocSim$, of locally simple categories. One can thus say that the notion of quasi locally simple category is equivalent to the notion of locally simple category and, in particular, that the notion of a category of not necessarily unitary regular algebras, is equivalent to that of a category of unitary regular algebras.

As the simple objects of a quasi locally simple category, $\mathbb{A}$, are of finite type, the category, $\mathbb{S}im\,\mathbb{A}$, of simple objects in $\mathbb{A}$, is the same as the category, $\mathbb{S}im\,\mathbb{A}_{tf}$ of simple objects in $\mathbb{A}_{tf}$. As a result, the categories $\mathbb{S}hv\mathbb{B}ool\mathbb{S}im\,\mathbb{A}$ and $\mathbb{S}hv\mathbb{L}oc\mathbb{B}ool\mathbb{S}im\,\mathbb{A}$, are equal respectively to the categories $\mathbb{S}hv\mathbb{B}ool\mathbb{S}im\,\mathbb{A}_{tf}$ and $\mathbb{S}hv\mathbb{L}oc\mathbb{B}ool\mathbb{S}im\,\mathbb{A}_{tf}$. By the theorem on the representation of locally simple categories by categories of Boolean sheaves of simple objects, one gets the equivalence $\mathbb{A}_{tf} \sim \mathbb{S}hv\mathbb{B}ool\mathbb{S}im\,\mathbb{A}_{tf}$, hence the equivalence : $\mathbb{A}^*_{tf} \sim (\mathbb{S}hv\mathbb{B}ool\mathbb{S}im\,\mathbb{A}_{tf})^*$ and finally the equivalence $\mathbb{A} \sim \mathbb{S}hv\mathbb{L}oc\mathbb{B}ool\mathbb{S}im\,\mathbb{A}$, that is to say, the representation of objects of $\mathbb{A}$ by locally Boolean sheaves of simple objects. In one direction, one associates to an object A of $\mathbb{A}$ the pair $(\mathrm{Spec}_{\mathrm{Max}}(A),\tilde{A})$ made up of the maximal spectrum of A, which is the set of maximal proper normal subobjects of A with the Stone topology on it, and the structural sheaf, $\tilde{A}$, on $\mathrm{Spec}_{\mathrm{Max}}(A)$, whose stalks are the simple quotient objects of A. Reversing this, one associates to a locally Boolean sheaf of simple objects in $\mathbb{A}$, the object of continuous global sections with compact support.

From the study of monadic morphisms of locally simple categories, one can infer the theory of monadic morphisms of quasi locally simple categories. A morphism of quasi locally simple categories is monadic if, and only if, it reflects isomorphisms. A monad $\mathbb{T} = (T,\eta,\mu)$ on a category $\mathbb{A}$ is quasi locally simple if the category $\mathbb{A}$ is quasi locally simple, the functor T preserves filtered colimits, and cokernels or finite products, and the natural transformation η is a semi-epimorphism. The functor forgetting structure, $U^{\mathbb{T}} : \mathbb{A}^{\mathbb{T}} \to \mathbb{A}$, is then a morphism of quasi locally simple categories. Monadic morphisms of quasi locally simple categories are characterised by the following "Beck theorem" : if $\mathbb{A}$ is a quasi locally simple category, a functor $U : \mathbb{X} \to \mathbb{A}$ is a monadic morphism of quasi locally simple categories if, and only if, the category $\mathbb{X}$ has a null object and generalised kernel pairs, the functor U has a left adjoint, preserves filtered colimits and creates regular quotient objects. If $\mathbb{A}$ is a quasi locally simple category, the full quasi locally simple subcategories of $\mathbb{A}$ are characterised by the property that they are reflexive full subcategories of $\mathbb{A}$ which are closed under filtered colimits and normal subobjects. If $\mathbb{K}$ is a full subcategory of $\mathbb{A}$ whose objects are finitely presentable, one can construct the universal quasi locally simple category in which the objects of $\mathbb{K}$ become null objects. It is the quotient category $\mathbb{A}/\mathbb{K}$. One shows that the quotient categories of $\mathbb{A}$ can be identified with the varieties of $\mathbb{A}$, to full quasi locally simple subcategories closed under subobjects, to full subcategories of $\mathbb{A}$ closed under products, subobjects and regular quotient objects or alternatively, to subcategories of $\mathbb{A}$ of the form $\mathbb{L}\mathrm{oc}\,\mathbb{S}$ where $\mathbb{S}$ is a full subcategory of $\mathbb{S}\mathrm{im}\,A$, closed under ultroproducts and subobjects, and where $\mathbb{L}\mathrm{oc}\,\mathbb{S}$ denotes the full subcategory of $\mathbb{A}$ whose objects are locally in $\mathbb{S}$, i.e. whose simple regular quotients are in $\mathbb{S}$. In particular, if $\mathbb{T}$ is an algebraic theory in the sense of Lawvere having two constants 0 and 1 and if $\mathbb{M}$ is a full subcategory of $\mathbb{A}\mathrm{lg}(\mathbb{T})$ closed under ultraproducts and subobjects, the category $\mathbb{S}\mathrm{hvLocBoolTsep}\,\mathbb{M}$ of totally separated locally Boolean sheaves of algebras from $\mathbb{M}$, is quasi locally simple and the categories of this form are precisely the quotient categories of the categories $\mathbf{S}$hvLocBoolTsepAlg($\mathbb{T}$).

CHAPTER 1

CATEGORIES OF BOOLEAN SHEAVES OF ALGEBRAS

1.0. LOCALLY FINITELY PRESENTABLE CATEGORIES.

1.0.0. Definitions. [16],[36]. An object, A, of a category $\mathbb{A}$ is said to be finitely presentable if the functor $\mathrm{Hom}_{\mathbb{A}}(A,-) : \mathbb{A} \to \mathbb{E}\mathrm{ns}$ preserves filtered colimits, and the category $\mathbb{A}$ is said to be locally finitely presentable if it is cocomplete and has a set, M, of finitely presentable objects, which is a proper generator in $\mathbb{A}$, i.e. M satisfies the following property : if f is any morphism of $\mathbb{A}$ and for all X in M, $\mathrm{Hom}_{\mathbb{A}}(X,f)$ is a bijection, then f is necessarily an isomorphism.

P. Gabriel and F. Ulmer in [16],[36] have given several representation theorems which show that these categories correspond to an axiomatic description of categories of sets with an algebraic structure.

1.1. LOCALLY INDECOMPOSABLE CATEGORIES.

1.1.0. Notations. Following A. Grothendieck and al. [21. 11.4.5], a sum, $(i_k : A_k \to \coprod_{k\in K} A_k)_{k\in K}$ of a family of objects of $\mathbb{A}$ is said to be disjoint if the injections i_k are monomorphisms and, for any pair (k,m) of distinct elements of K, the fibred product of (i_k, i_m) is the initial object of $\mathbb{A}$. Dually, a product $(p_k : \prod_{k\in K} A_k \to A_k)_{k\in K}$ of a family of objects of $\mathbb{A}$ is said to be codisjoint if the projections p_k, are epimorphisms and, for any pair (k,m) of distincts elements of K, the amalgamated sum of (p_k, p_m) is the final object of $\mathbb{A}$. If $(f : A \to B,\ g : A \to C)$ is a pair of morphisms with the same source, having as amalgamated sum the pair $(m : B \to D,\ n : C \to D)$, the morphism n is called the direct image of f by g. Dually, if $(f : A \to B,\ g : C \to B)$ is a pair of morphisms with the same target, having as a fibred product the pair $(m : D \to A,\ n : D \to C)$, the morphism n is called the inverse image of f by g. Following A. Grothendieck and al. [21, 1.2.5], a sum $(i_k : A_k \to \coprod_{k\in K} A_k)_{k\in K}$ is said to be universal if for any morphism $f : B \to \coprod_{k\in K} A_k$, the family of morphisms

$(j_k : B_k \to B)_{k\in K}$, which are the inverse images of the injections i_k by f, expresses B as a sum. In a dual way, a product

$(p_k : \prod_{k\in K} A_k \to A_k)_{k\in K}$ is said to be couniversal if, for any morphism $f : \prod_{k\in K} A_k \to B$, the family of morphisms $(q_k : B \to B_k)_{k\in K}$ which are the direct images of the projections p_k by f, expresses B as a product. In particular, the final object 1, product of the empty family of objects, is couniversal if, and only if, it is strict, that is to say that any morphism with source 1 is an isomorphism.

1.1.1. Definition. [15]. A category is locally indecomposable if:

(1) it is locally finitely presentable,

(2) the product of its initial object with itself is finitely presentable,

(3) its finite products are codisjoint and couniversal.

The initial object is denoted Z and its product with itself is denoted $(\pi : Z^2 \to Z, \pi' : Z^2 \to Z)$. We note that finite products are codisjoint and couniversal if, and only if, the final object 1, is strict and the products of two objects are codisjoint and couniversal.

1.2. EXAMPLES OF LOCALLY INDECOMPOSABLE CATEGORIES.

All the rings considered are unitary.

1.2.1. $\mathbb{R}$ngc : category of commutative rings.

It is locally finitely presentable. The ring $\mathbb{Z}$ is the initial object and its square $\mathbb{Z}^2 \simeq \mathbb{Z}[X]/(X^2-X)$ is of finite presentation. The ring $\{0\}$ is a strict final object. Let $A,B,C \in \mathbb{R}$ngc and $f : A\times B \to C$. Set $e = f(0,1)$. The element e is idempotent and $1-e = f(1,0)$. The projection $A\times B \to A$ is isomorphic to the quotient of $A\times B$ by the ideal generated by $(0,1)$, thus its direct image by f is isomorphic to the quotient of C by the ideal generated by e. Similarly, the direct image of the projection $A\times B \to B$ by f is isomorphic to the quotient of C by the ideal generated by $1-e$. However, the pair of quotient rings $(C \to C/(e),\ C \to C/(1-e))$ is a product. Therefore finite products are couniversal in $\mathbb{R}$ngc. They are codisjoint, since projections are surjective and the amalgamated sum of projections $(A\times B \to A,\ A\times B \to B)$ is isomorphic to the quotient of $A\times B$ by the ideal generated by $\{(0,1),(1,0)\}$, that is to $\{0\}$.

1.2.2. <u>Rngc(p)</u> : category of commutative rings of characteristic p with p a prime.

It is the full sub-category of Rngc having as objects the rings that satisfy $p.1 = 0$.

1.2.3. <u>p-Rngc</u> : category of p-rings [32].

1.2.4. <u>p^k-Rngc</u> : category of p^k-rings [32].

1.2.5. <u>RngcRed</u> : reduced commutative rings

1.2.6. <u>RngcReg</u> : regular commutative rings [34].

1.2.7. <u>RngcFormRl</u> : formally real commutative rings, i.e. satisfying the axiom : $1 + X_1^2 + \dots + X_n^2$ is invertible [27].

1.2.8. <u>RngcO</u> : orderable commutative rings with positive squares, i.e., satisfying the axiom : $X_1^2 + \dots + X_n^2 = 0 \Rightarrow X_1^2 = 0$ [7].

1.2.9. <u>RngcRedO</u> : reduced orderable commutative rings with positive squares, i.e., satisfying the axiom : $X_1^2 + \dots + X_n^2 = 0 \Rightarrow X_1=0$, [7].

1.2.10. <u>RngcOAnnConv</u> : orderable commutative rings with positive squares and convex annihilators, i.e. satisfying the axiom : $(X_1^2 + \dots + X_n^2)Y = 0 \Rightarrow X_1^2Y = 0$, [7].

1.2.11. <u>RngcRepDom</u> : commutative rings representable by domains [25].

1.2.12. <u>RngcBa</u> : commutative Baer rings and homomorphisms preserving the idempotent associated to each element [22].

1.2.13. <u>Bool</u> : Boolean Algebras

1.2.14. <u>RngSym</u> : symmetric rings [29].

1.2.15. <u>RngRed</u> : reduced rings.

1.2.16. <u>RngStReg</u> : strongly regular rings [1]

1.2.17. <u>RngcDif</u> : differential rings and differential homomorphisms [28].

1.2.18. <u>RngcDif(p)</u> : differential rings of prime characteristic p, i.e. satisfying $p.1 = 0$ [28].

1.2.19. <u>RngcDifRed</u> : reduced differential rings.

1.2.20. <u>RngcDifReg</u> : regular differential rings.

1.2.21. <u>RngcDifRegPerf(p)</u> : regular differential rings, differentially perfect, of prime characteristic p [28].

1.2.22. $\mathbb{R}$ngc$\mathbb{R}$eg$\mathbb{O}$rd : ordered regular commutative rings with positive squares and increasing homomorphisms.

1.2.23. $\mathbb{R}$ngc$\mathbb{R}$eg$\mathbb{L}$at : regular, lattice-ordered, commutative rings with positive squares.

1.2.24. $\mathbb{R}$ngc$\mathbb{O}$rd : ordered commutative rings with positive squares [7].

1.2.25. $\mathbb{R}$ngc$\mathbb{O}$rd$\mathbb{R}$ed : ordered reduced commutative rings with positive squares [7].

1.2.26. $\mathbb{R}$ngc$\mathbb{O}$rd$\mathbb{A}$nn$\mathbb{C}$onv : ordered commutative rings with positive squares and convex annihilators [7].

1.2.27. $\mathbb{R}$ngc$\mathbb{L}$at : lattice-ordered commutative rings with positive squares and homomorphisms preserving finite meets and joins [24].

1.2.28. $\mathbb{R}$ngc$\mathbb{S}$t$\mathbb{L}$at : strongly lattice-ordered commutative rings [24].

1.2.29. $\mathbb{R}$ngc$\mathbb{S}$t$\mathbb{L}$at$\mathbb{P}$roj : strongly lattice-ordered projectable rings [24].

1.2.30. $\mathbb{R}$ngc$\mathbb{S}$t$\mathbb{L}$at$\mathbb{R}$ep$\mathbb{D}$om : strongly lattice-ordered rings representable by totally ordered domains [25].

1.2.31. $\mathbb{R}$ngc$\mathbb{S}$t$\mathbb{L}$at$\mathbb{B}$a : strongly lattice-ordered Baer rings.

1.2.32. $\mathbb{A}$lg(K) : commutative algebras over a commutative ring K.

1.2.33. $\mathbb{A}$lg$\mathbb{D}$if(K) : differential algebras over K.

1.2.34. $\mathbb{L}$atDist : bounded distributive lattices.

1.2.35. $\mathbb{H}$eyt : Heyting algebras [23].

1.3. THE DIRECT FACTOR COCLASSIFIER.

1.3.0. Quotient objects of an object.

Let $\mathbb{A}$ be a locally indecomposable category. The category $\mathbb{A}$ has few quotient objects [16], i.e. it is co-well-powered [33, p.126]. For each object A of $\mathbb{A}$, one denotes by $Q(A)$, the set of quotient objects of A and one identifies, by abuse of language, a quotient object with one of its representatives [33, p.122]. The set $Q(A)$ is canonically ordered by $\delta : A \to B \leqslant \mu : A \to C$ if, and only if, there exists a morphism $n : C \to B$ satisfying $n\mu = \delta$, [33, p.122]. Since $\mathbb{A}$

is cocomplete, $Q(A)$ is a complete lattice, whose meets are called cointersections and whose joins, counions, the maximum element is $1_A : A \to A$ and the minimum element is $0_A : A \to 1$.

A regular epimorphism is a morphism $q : A \to Q$ which is a coequaliser of a pair of morphisms $(m,n) : X \rightrightarrows A$, [2, 1.1]. It represents a regular quotient object of A. One denotes by $R(A)$, the set of regular quotient objects of A. It is a complete sub-inf-semi-lattice of $Q(A)$, and is therefore a complete lattice. A regular quotient object $q : A \to Q$ is finitely generated if q is coequaliser of a pair of morphisms $(m,n) : X \rightrightarrows A$, where X is a finitely generated object of $\mathbb{A}$ [16]. As any finitely generated object of $\mathbb{A}$ is a quotient object of a finitely presentable object, this is the same as requiring that X be finitely presentable. These quotient objects are precisely the cocompact elements of $R(A)$ [18, section 9, definition 1.2.]. As any object of $\mathbb{A}$ is a filtered colimit of finitely presentable objects, any regular quotient object of A is a cointersection of cocompact regular quotient objects of A, so the lattice $R(A)$ is cocompactly cogenerated [18]. If $(\delta : A \to B,\ \delta' : A \to B')$ is a product in $\mathbb{A}$, the two morphisms δ and δ' are epimorphisms, since finite products are codisjoint ; they therefore define quotient objects of A.

1.3.1. Definition. A direct factor of A is a quotient object of A of the form $\delta : A \to B$ such that there exists a morphism $\delta' : A \to B'$, forming with δ a product $(\delta : A \to B,\ \delta' : A \to B')$.

The direct factors of A form a set, $\Delta(A)$, ordered by the order induced from that of $Q(A)$.

1.3.2. Proposition. Direct factors are precisely the complemented quotient objects. Isomorphisms, and morphisms of target 1 are direct factors. Direct factors are couniversal and stable under composition and amalgamated sums.

Proof : Let $(\delta : A \to B,\ \delta' : A \to B')$ be a product. The relation $\delta \wedge \delta' = 0_A : A \to 1$ results from the fact that finite products are codisjoint. If $\mu : A \to C$ is a quotient object such that $\delta \leqslant \mu$ and $\delta' \leqslant \mu$, there exist two morphisms $m : C \to B$ and $m' : C \to B'$ satisfying $m\mu = \delta$ and $m'\mu = \delta'$, and thus there will be a morphisms $n : C \to A$ satisfying $\delta n = m$ and $\delta' n = m'$. The morphism n, is inverse to μ and consequently μ is an isomorphism. Therefore $\delta \vee \delta' = 1_A$ and δ' is complementary to δ in $Q(A)$. Conservely, let us consider two quotient objects, $\delta : A \to B$ and $\delta' : A \to B'$, which are complementary in $Q(A)$. Let us denote by $(p : B\times B' \to B,\ p' : B\times B' \to B')$, the product of B and B', and by $(\delta,\delta') : A \to B\times B'$, the resulting canonical morphism. The amalgamated sum of $(\delta : A \to B,\ \delta : A \to B)$ is $(1_B : B \to B,\ 1_B : B \to B)$ and the amalgamated sum of $(\delta' : A \to B',\ \delta : A \to B)$ is $(0_{B'} : B' \to 1,\ 0_B : B \to 1)$. Since finite products are couniversal, the amalgamated sum of $((\delta,\delta') : A \to B\times B',\ \delta : A \to B)$ is $(p : B\times B' \to B,\ 1_B : B \to B)$. Similarly, the amalgamated sum of $((\delta,\delta') : A \to B\times B',\ \delta' : A \to B')$ is $(p' : B\times B' \to B',\ 1_{B'} : B' \to B')$. Consequently, the amalgamated sum of $((\delta,\delta') : A \to B\times B',\ (\delta,\delta') : A \to B\times B')$ is $(1_{B\times B}, 1_{B\times B})$. The morphism $(\delta,\delta') : A \to B\times B'$ is therefore an epimorphism. The relations, $\delta \leqslant (\delta,\delta')$ and $\delta' \leqslant (\delta,\delta')$, then imply $(\delta,\delta') = 1_A$. The morphism (δ,δ') is therefore an isomorphism and the pair $(\delta : A \to B,\ \delta' : A \to B')$ is a product. This shows that complementary quotient objects are direct factors. If $\delta : A \to B$ is an isomorphism, the pair $(\delta : A \to B,\ 0_A : A \to 1)$ is a product and it follows that δ and 0_A are direct factors. Since finite products are couniversal, so are direct factors. If $\delta : A \to B$ and $\mu : B \to C$ are two direct factors, there exist $\delta' : A \to B'$ and $\mu' : B \to C'$ such that $(\delta : A \to B,\ \delta' : A \to B')$ and $(\mu : B \to C,\ \mu' : B \to C')$ are products. Then $(\mu\delta : A \to C,\ (\mu'\delta,\delta') : A \to C'\times B')$ is a product and $\mu\delta$ is a direct factor. If $\delta : A \to B$ and $\mu : A \to C$ are two direct factors and $(r : B \to D,\ s : C \to S)$ is their amalgamated sum, then r is a direct factor, as is $r\delta$. ■

1.3.3. <u>Proposition</u>. <u>The set</u> $\Delta(A)$ <u>of direct factors of</u> A <u>is a Boolean algebra in which the complement of</u> $\delta : A \to B$ <u>is the unique direct factor</u> $\delta' : A \to B'$ <u>forming with</u> δ, <u>a product</u> $(\delta : A \to B,\ \delta' : A \to B')$.

<u>Proof</u> : The direct factor 1_A is the maximum and the direct factor 0_A is the minimum. If $\delta : A \to B$ and $\mu : A \to C$ are two direct factors, then their meet in $Q(A)$ is represented by the amalgamated sum of (δ,μ) ; it is a direct factor which is the meet of δ and μ in $\Delta(A)$. Any direct factor $\delta : A \to B$ has a complementary direct factor, $\delta' : A \to B'$, such (δ,δ') is a product. Let us show that δ' is also pseudo-complementary to δ in $Q(A)$ [18, p.58]. Let $\mu : A \to C$ in $Q(A)$ be such that $\mu \wedge \delta = 0_A$. The direct image of δ by μ is $0_C : C \to 1$ and the direct image of δ' by μ is a direct factor, $\nu : C \to C'$, which with 0_A gives a product $(\nu,0_A)$. Consequently, ν is an isomorphism and $\mu \leqslant \delta'$. In particular, this implies the

uniquenes of δ'. For each pair (δ,μ) of direct factors of A, we define the direct factor $\delta \vee \mu$ by $\delta \vee \mu = (\delta' \wedge \mu')'$. Let us prove that it is the join of δ and μ in $\Delta(A)$ (cf. [18, theorem 4, section 6]). To start with $\delta \leqslant \delta \vee \mu$, since $\delta \wedge \delta' \wedge \mu' = 0_A$; similarly $\mu \leqslant \delta \vee \mu$. Next, if $\nu \in \Delta(A)$ is such that $\delta \leqslant \nu$, then $\delta \wedge \nu' = 0_A$ and $\mu \wedge \nu' = 0_A$, therefore $\nu' \leqslant \delta'$ and $\nu' \leqslant \mu'$. It follows that $\nu' \leqslant \delta' \wedge \mu'$, so $\nu' \wedge (\delta' \wedge \mu')' = 0_A$, and $(\delta' \wedge \mu')' \leqslant \nu$, that is $\delta \vee \mu \leqslant \nu$. It remains to prove distributivity. Let $\delta,\mu,\ \nu \in \Delta(A)$. One has $\delta \wedge \mu \wedge (\delta \wedge \mu') \wedge (\delta \wedge \nu)' = 0_A$ and $\delta \wedge \nu \wedge (\delta \wedge \mu') \wedge (\delta \wedge \nu)' = 0_A$, so $\mu \leqslant (\delta \wedge (\delta \wedge \mu)' \wedge (\delta \wedge \nu)')'$ and $\nu \leqslant (\delta \wedge (\delta \wedge \mu)' \wedge (\delta \wedge \nu)')'$ thus $\mu \vee \nu \leqslant (\delta \wedge (\delta \wedge \mu)' \wedge (\delta \wedge \nu)')'$, which implies that $(\mu \vee \nu) \wedge \delta \wedge (\delta \wedge \mu)' \wedge (\delta \wedge \nu)' = 0_A$, therefore $(\mu \vee \nu) \wedge \delta \leqslant (\delta \wedge \mu) \vee (\delta \wedge \nu)$. ∎

1.3.4. Theorem. In a locally indecomposable category, the square Z^2 of the initial object is a coclassifying object for direct factors, i.e., for any direct factor $\delta : A \to B$, there exists a unique morphism $f : Z^2 \to A$ such that the direct image of the projection $\pi : Z^2 \to Z$ by f is δ.

Proof : Let $\delta : A \to B$ be a direct factor and $\delta' : A \to B'$ its complement and let us write $i_B : Z \to B$, $i_{B'} : Z \to B'$ for the unique morphisms. If $f : Z^2 \to A$ is a morphism such that the direct image of π by f is δ, then the direct image of π' by f is δ', thus f is necessarily the morphism $f = i_B \times i_{B'} : Z^2 \to B \times B' = A$. If μ is the direct image of π by $i_B \times i_B$, and μ' is the direct image of π' by $i_B \times i_{B'}$, one has : $\delta \leqslant \mu$ and $\delta' \leqslant \mu'$, which implies $\mu = \delta$. ∎

1.3.5. The functor $\Delta : \mathbb{A} \to \mathbb{B}\mathrm{ool}$.

For every morphism $f : A \to B$, of $\mathbb{A}$, let us write $\Delta(f) : \Delta(A) \to \Delta(B)$ for the mapping which assigns to $\delta \in \Delta(A)$ its direct image by f. Since the direct image by f preserves the minimum quotient object, amalgamated sums and finite products, the mapping $\Delta(f)$ preserves finite meets and complements and, consequently, is a Boolean algebra homomorphism. We thus obtain a functor, $\Delta : \mathbb{A} \to \mathbb{B}\mathrm{ool}$. If one writes $U_o : \mathbb{B}\mathrm{ool} \to \mathbb{E}\mathrm{ns}$ for the underlying set functor, the functor, $U_o\Delta : \mathbb{A} \to \mathbb{E}\mathrm{ns}$, can be represented by the object Z^2 and the pair (Z^2,π) is a representing pair. The functor $U_o\Delta$ therefore preserves limits and, as the object Z^2 is finitely presentable, it also preserves filtered colimits. As the functor U_o creates limits and filtered colimits, the functor Δ preserve limits and filtered colimits.

1.4. FINITELY PRESENTABLE OBJECTS.

1.4.0. Proposition. Finitely presentable objects in a locally indecomposable category are stable under finite products.

Proof :

(a) Let $\mathbb{A}$ be a locally indecomposable category. The final Boolean algebra $\{0\}$, is finitely presentable in $\mathbb{Bool}$ and the functor $\Delta : \mathbb{A} \to \mathbb{Bool}$ preserves and reflects final objects. It follows that the identity morphism $\{0\} \to \Delta(1)$ is a universal morphism from 0 to Δ and, that for any filtered colimit $B = \varinjlim_{i\in\mathbb{I}} B_i$ in $\mathbb{A}$, one has $\mathrm{Hom}_{\mathbb{A}}(1,\varinjlim_{i\in\mathbb{I}} B_i) \simeq \mathrm{Hom}_{\mathbb{Bool}}(\{0\},\Delta(\varinjlim_{i\in\mathbb{I}} B_i) \simeq \mathrm{Hom}_{\mathbb{Bool}}(\{0\},\varinjlim_{i\in\mathbb{I}}\Delta(B_i))\simeq \varinjlim_{i\in\mathbb{I}} \mathrm{Hom}_{\mathbb{Bool}}(\{0\},\Delta(B_i)) \simeq \varinjlim_{i\in\mathbb{I}} \mathrm{Hom}_{\mathbb{A}}(1,B_i)$; the final object 1 is thus finitely presentable in $\mathbb{A}$.

(b) Let $Z^2/\mathbb{A}$ be the category of objects of $\mathbb{A}$ under Z^2. We are going to prove that if (A,f) is a finitely presentable object of $Z^2/\mathbb{A}$, then A is a finitely presentable object of $\mathbb{A}$. Let $(\tau_i : B_i \to B)_{i\in\mathbb{I}}$ be a filtered colimit in $\mathbb{A}$, and let $g : A \to B$. As Z^2 is finitely presentable in $\mathbb{A}$, there exist an object i_o of $\mathbb{I}$ and a morphism $f_o : Z^2 \to B_{i_o}$ such that $\tau_{i_o} f_o = gf$. Let us note that the category $i_o/\mathbb{I}$ of objects of $\mathbb{I}$ under i_o is filtered and that the target functor $\beta : i_o/\mathbb{I} \to \mathbb{I}$ is cofinal. It follows that the inductive cone $(\iota_{\beta(u)} : B_{\beta(u)} \to B)_{u\in i_o/\mathbb{I}}$ is a filtered colimit in $\mathbb{A}$. If, for each object u of $i_o/\mathbb{I}$, we write $f_u = B_u f_o$, one obtains a filtered colimit $(\iota_{\beta(u)} : (B_{\beta(u)},f_u) \to (B,gf))_{u\in i_o/\mathbb{I}}$ in the category $Z^1/\mathbb{A}$, as well as a morphism $g : (A,f) \to (B,gf)$. Since the object (A,f) is finitely presentable in $Z^2/\mathbb{A}$, there exist an object $u \in i_o/\mathbb{I}$ and a morphism $g_u : (A,f) \to (B_{\beta(u)},f_u)$ satisfying $\iota_{\beta(u)} g_u = g$. The morphism $g : A \to B$ can thus be factorised through the injection $i_{\beta(u)}$. Let us consider an object j of $\mathbb{I}$ and a morphism $g_j : A \to B_j$ satisfying $\iota_j g_j = g$. Since Z^2 is finitely presentable, $i_j g_j f = gf = \iota_{\beta(u)} g_u f$ implies the existence of two morphisms $v : \beta(u) \to k$ and $w : j \to k$ of $\mathbb{I}$ satisfying the relations $B_w g_j f = B_v g_u f = f_{vu}$. In this way, we obtain two morphisms $B_w g_j : (A,f) \to (B_k,f_{vu})$ and $B_v g_u : (A,f) \to (B_k,f_{vu})$ of $Z^2/\mathbb{A}$ coequalised by the morphism $i_k : (B_k,f_{vu}) \to (B,gf)$. Since (A,f) is finitely presentable, there exists a morphism $f : (k,vu) \to (m,tvu)$ of $i_o/\mathbb{I}$ such that $B_t B_w g_j = B_j B_v g_u$, therefore $B_{tw} g_j = B_{tv} g_u$. This completes the proof that the inductive cone

$(\mathrm{Hom}_{\mathbb{A}}(A,\iota_i) : \mathrm{Hom}_{\mathbb{A}}(A,B_i) \to \mathrm{Hom}_{\mathbb{A}}(A,B))_{i\in\mathbb{I}}$ is a filtered colimit in $\mathbb{E}ns$ and consequently that the object A is finitely presentable in $\mathbb{A}$.

(c) Let $P : \mathbb{A}\times\mathbb{A} \to Z^2/\!/\mathbb{A}$ be the functor defined on objects by $P(A,B) = (A\times B,\ i_A\times i_B)$ where $i_A : Z \to A$, $i_B : Z \to B$ are the unique morphisms, and on morphisms by $P(f,g) = f\times g$. Let $Q : Z^2/\!/\mathbb{A} \to \mathbb{A}\times\mathbb{A}$ be the functor defined in the following way : if (A,f) is an object of $Z^2/\mathbb{A}$, $\delta : A \to B$ (resp. $\delta' : A \to B'$) will denote a direct image of π (resp. π') by f and we will set $Q(A,f) = (B,B')$; if $g : (A,f) \to (M,m)$ is a morphism of $Z^2/\mathbb{A}$, $\mu : M \to N$ (resp. $\mu' : M \to N'$) will denote the direct image of π (resp. π') by m so that $Q(M,m) = (N,N')$, writing (h,μ) for the amalgamated sum of (δ,g), and (h',μ') for the amalgamated sum of (δ',g), we will set $Q(g) = (h,h')$. It is immediate that the functors P and Q establish an equivalence of categories (cf. Theorem 1.3.4.).

(d) If A and B are two finitely presentable objects of $\mathbb{A}$, the pair (A,B) is a finitely presentable object of $\mathbb{A}\times\mathbb{A}$, and consequently, the object $P(A,B) = (A\times B,\ i_A\times i_B)$ is finitely presentable in $Z^2/\mathbb{A}$, which, in turn, implies that $A\times B$ is finitely presentable in $\mathbb{A}$. ∎

1.5. <u>INDECOMPOSABLE OBJECTS</u>.

1.5.0. <u>Definition</u>. An object of $\mathbb{A}$ is <u>indecomposable</u> if it has exactly two direct factors.

1.5.1. <u>Proposition</u>. <u>For an object</u> A <u>of</u> $\mathbb{A}$, <u>the following statements are equivalent</u> :

(0) A <u>is indecomposable</u>,

(1) <u>the Boolean algebra</u> $\Delta(A)$ <u>is isomorphic to the Boolean algebra</u> $2 = \{0,1\}$,

(2) $A \neq 1$ <u>and any direct factor of</u> A <u>is an isomorphism or a morphism with target</u> 1,

(3) $A \neq 1$ <u>and any product</u> $(p : A \to B,\ q : A \to C)$ <u>is such that</u> p <u>or</u> q <u>is an isomorphism,</u>

(4) <u>the functor</u> $\mathrm{Hom}_{\mathbb{A}}(-,A) : \mathbb{A}^{op} \to \mathbb{E}ns$ <u>preserves finite sums</u>,

(5) $A \neq 1$ <u>and any morphism</u> $Z^2 \to A$ <u>can be factorised through</u> $\pi : Z^2 \to Z$ <u>or</u> $\pi' : Z^2 \to Z$.

<u>Proof</u> : The equivalence (0) <=> (1) <=> (2) <=> (3) are

immediate.

(3) => (4). The functor $\mathrm{Hom}_{\mathbb{A}}(-,A)$ preserves the sum of the empty family since $A \neq 1$ implies that $\mathrm{Hom}_{\mathbb{A}}(1,A) = \emptyset$. Let $(p : B\times C \to B,\ q : B\times C \to C)$ be a product. The function $\langle \mathrm{Hom}_{\mathbb{A}}(p,A), \mathrm{Hom}_{\mathbb{A}}(q,A)\rangle : \mathrm{Hom}_{\mathbb{A}}(B,A) \coprod \mathrm{Hom}_{\mathbb{A}}(C,A) \to \mathrm{Hom}_{\mathbb{A}}(B\times C,A)$ is injective, since finite products are codisjoint and $A \neq 1$. It is also surjective as, for $f : B\times C \to A$, the direct image $\delta : A \to D$ of p by f and the direct image $\delta' : A \to D'$ of q by f form a product $(\delta : A \to D, \delta' : A \to D')$ and consequently one of the morphisms δ or δ' is an isomorphism, which implies that f is factorisable through p or q.

(4) => (5). Since $\mathrm{Hom}_{\mathbb{A}}(1,A) = \emptyset$, $A \neq 1$ and the two morphisms $i_A\pi : Z^2 \to A$ and $i_A\pi' : Z^2 \to A$ are distinct and they are the only morphisms $Z^2 \to A$, since $\mathrm{Hom}_{\mathbb{A}}(Z^2,A) \to \mathrm{Hom}_{\mathbb{A}}(Z,A) \coprod \mathrm{Hom}_{\mathbb{A}}(Z,A)$ has exactly two elements.

(5) => (0). $\Delta(A) \simeq \mathrm{Hom}_{\mathbb{A}}(Z^2,A)$ has exactly two elements. ∎

The category $\mathbb{I}\mathrm{nd}\,\mathbb{A}$ of indecomposable objects of $\mathbb{A}$ is the full subcategory of $\mathbb{A}$ whose objects are the indecomposable objects.

1.5.2. Proposition. The subcategory $\mathbb{I}\mathrm{nd}\,\mathbb{A}$ is closed, in $\mathbb{A}$, under subobjects, filtered colimits, and ultraproducts.

Proof : The full subcategory of $\mathbb{B}\mathrm{ool}$ having as its objects the Boolean algebras isomorphic to $2 = \{0,1\}$, is closed under subobjects, filtered colimits and ultraproducts. The category $\mathbb{I}\mathrm{nd}\,\mathbb{A}$ is the inverse image of this subcategory by the functor $\Delta : \mathbb{A} \to \mathbb{B}\mathrm{ool}$. However, the functor Δ preserves limits and filtered colimits and consequently monomorphisms and ultraproducts. Thus $\mathbb{I}\mathrm{nd}\,\mathbb{A}$ is closed in $\mathbb{A}$ for subobjects, filtered colimits and ultraproducts. ∎

1.6. INDECOMPOSABLE COMPONENTS OF AN OBJECT.

Let A be an object of $\mathbb{A}$.

1.6.0. Definitions. (1) An indecomposable quotient object of A is a quotient object $q : A \to Q$ of A such that the object Q is indecomposable.

(2) An indecomposable component of A is a maximal indecomposable quotient object of A.

1.6.1. Proposition. For any ultrafilter, Φ , of the Boolean algebra $\Delta(A)$, the quotient object $\gamma_\Phi : A \to A_\Phi$, meet of the direct factors of A belonging to Φ, is an indecomposable component of A. The mapping $\Phi \to \gamma_\Phi$ is a bijection from the set of ultrafilters of $\Delta(A)$ onto the set of indecomposable components of A.

Proof : (a) For any direct factor $\delta : A \to B$ of A, let us write $\beta(\delta) = B$, the target object. Equipped with the order induced by that of $\Delta(A)$, an ultrafilter Φ of $\Delta(A)$ is a cofiltered ordered set, which defines a filtered diagram $(\beta(\delta))_{\delta\in\Phi}$ of $\mathbb{A}$ indexed by Φ^{op}, the colimit of which will be written $(i_\delta : \beta(\delta) \to A_\Phi)_{\delta\in\Phi}$. The morphism $i_{1_A} : A \to A_\Phi$ represents a quotient object of A, written γ_Φ, which is the meet of the direct factors of A belonging to Φ. For each $\delta \in \Delta(A)$, the Boolean algebra $\Delta(\beta(\delta))$ is isomorphic to the quotient Boolean algebra, $\Delta(A)/(\delta)$ of $\Delta(A)$ by the principal filter generated by δ. It follows that

$$\Delta(A_\Phi) = \Delta(\varinjlim_{\delta\in\Phi}\beta(\delta)) \simeq \varinjlim_{\delta\in\Phi}\Delta(\beta(\delta)) \simeq \varinjlim_{\delta\in\Phi}\Delta(A)/(\delta) \simeq \Delta(A)/\Phi \simeq 2.$$

The object A_Φ is therefore indecomposable. Let us show that $\gamma_\Phi : A \to A_\Phi$ is an indecomposable component of A. Let $q : A \to Q$ be an indecomposable quotient object of A such that $\gamma_\Phi \leq q$, that is to say, such that γ_Φ can be factorised through q. For each $\delta \in \Phi$, the morphism q factorises through δ or through δ'. However q cannot be factorised through δ', or γ_Φ would be factorisable through δ' and δ, which is contrary to the fact that A_Φ is indecomposable. Thus q can be factorised through all the morphisms $\delta \in \Phi$ and it follows that $q \leq \gamma_\Phi$, so $\gamma_\Phi = q$.

(b) Let Φ be an ultrafilter of $\Delta(A)$ and $\delta \in \Delta(A)$. If $\delta \in \Phi$, then δ factorises γ_Φ. Conversely, if δ factorises δ_Φ, then δ' does not factorise γ_Φ, so $\delta' \in \Phi$ and it follows that $\delta \in \Phi$. Thus, we have $\delta \in \Phi \iff \delta$ factorises γ_Φ. It follows that if Φ and Φ' are two ultrafilters of $\Delta(A)$ such that $\gamma_\Phi = \gamma_{\Phi'}$, then $\delta \in \Phi \iff \delta$ factorises $\gamma_\Phi = \gamma_{\Phi'} \iff \delta \in \Phi$, thus $\Phi = \Phi'$. The mapping, $\Phi \to \gamma_\Phi$ is therefore injective. Let $\gamma : A \to C$ be an indecomposable component of A. The set Φ of direct factors of A which factorise γ, is the inverse image of the direct factor 1_C by the homomorphism $\Delta(\gamma) : \Delta(A) \to \Delta(C)$. It is therefore an ultrafilter of $\Delta(A)$. The indecomposable component γ_Φ is therefore greater than, or equal to, γ, and so they must be equal. The mapping $\Phi \to \gamma_\Phi$ is therefore surjective.

∎

1.7. THE INDECOMPOSABLE SPECTRUM OF AN OBJECT AND ITS STRUCTURAL SHEAF.

Recall, [10] and [14], that a full subcategory $\mathbb{X}$ of $\mathbb{A}$ is multireflexive if, for any object, A, of $\mathbb{A}$, there exists a universal family of morphisms from A towards $\mathbb{X}$, that is a family $(\eta_i : A \to X_i)_{i\in I}$ of morphisms with source A, whose target is in $\mathbb{X}$ and such that any morphism $g : A \to X$ from A to an object X of $\mathbb{X}$ is factorised in a unique way through a unique morphism of the family. Let us also recall [11] that a category is said to be locally $\aleph_0$-multipresentable if it is finitely multicocomplete, has filtered colimits and a proper generating set formed of finitely presentable objects.

1.7.0. Proposition. The category $\mathbb{I}\text{nd}\,\mathbb{A}$ of indecomposable objects of $\mathbb{A}$ is a multireflexive subcategory of $\mathbb{A}$ and it is locally $\aleph_0$-multipresentable.

Proof : Let us show that the family of indecomposable components of an object A of $\mathbb{A}$ is a universal family of morphisms of A towards $\mathbb{I}\text{nd}\,\mathbb{A}$. Let $f : A \to K$ with $K \in \mathbb{I}\text{nd}\,\mathbb{A}$. The set Φ of direct factors of A which factorise f is an ultrafilter of $\Delta(A)$ (cf. (b) proof. 1.6.1.) and the indecomposable component $\gamma_\Phi : A \to A_\Phi$ factorises f in a unique way, since γ_Φ is an epimorphism. If Φ' is an ultrafilter of $\Delta(A)$ such that $\gamma_{\Phi'}$ factorises f, then $\delta \in \Phi' \Rightarrow \delta$ factorises $\gamma_{\Phi'} \Rightarrow \delta$ factorises $f \Rightarrow \delta \in \Phi$ and consequently $\Phi' \subset \Phi$, so $\Phi' = \Phi$. As a result, $\mathbb{I}\text{nd}\,\mathbb{A}$ is a full, multireflective subcategory of $\mathbb{A}$ [10, 3.3.1.]. Since it is closed in $\mathbb{A}$ under filtered colimits, it is a locally $\aleph_0$-multipresentable category [13, Prop. 3.3.]. ■

Following the general notation for multiadjoints [14, section 1], for any object A of $\mathbb{A}$, the set of ultrafilters of the Boolean algebra $\Delta(A)$, is denoted $\text{Spec}_{\text{Ind}}(A)$ and is called the indecomposable spectrum of A, and for any morphism $f : A \to B$, the homomorphism of Boolean algebras $\Delta(f) : \Delta(A) \to \Delta(B)$ defines the function $\text{Spec}_{\text{Ind}}(f) : \text{Spec}_{\text{Ind}}(B) \to \text{Spec}_{\text{Ind}}(A)$, which assignes to an ultrafilter of $\Delta(B)$, its inverse image by $\Delta(f)$. The spectral topology on $\text{Spec}_{\text{Ind}}(A)$ is the Stone topology [18, section 11] defined in the following way. Let us write $D : \Delta(A) \to \mathcal{P}(\text{Spec}_{\text{Ind}}(A))$ for the function defined by $D(\delta) = \{\Phi \in \text{Spec}_{\text{Ind}}(A) : \delta \in \Phi\}$. It is strictly increasing and induces an isomorphism between the Boolean algebra $\Delta(A)$ and its image denoted $\mathcal{D}(\text{Spec}_{\text{Ind}}(A))$. The set $\mathcal{D}(\text{Spec}_{\text{Ind}}(A))$ is a base of open sets for the Stone topology on $\text{Spec}_{\text{Ind}}(A)$. For every morphism $f : A \to B$, the function, $\mathrm{Spec}_{\mathrm{Ind}}(f) : \mathrm{Spec}_{\mathrm{Ind}}(B) \to \mathrm{Spec}_{\mathrm{Ind}}(A)$ is continuous. The ordered set $\mathcal{D}(\mathrm{Spec}_{\mathrm{Ind}}(A))$, considered as a category, canonically has the structure of a site, [21, 11.1.15], the topology, T, of which has, as covering families, the families $(u_i \to u)_{i\in I}$ such that $u = \bigcup_{i\in I} u_i$. The functor $\tilde{A} : \mathcal{D}(\mathrm{Spec}_{\mathrm{Ind}}(A))^{op} \to \mathbb{A}$ is defined by $\tilde{A}(D(\delta : A \to B)) = B$ and $\tilde{A}(D(\delta) \subset D(\mu)) = m$ for two direct factors $\delta : A \to B$ and $\mu : A \to C$ satisfying $m\mu = \delta$.

1.7.1. Theorem. The functor $\tilde{A} : \mathcal{D}(\mathrm{Spec}_{\mathrm{Ind}}(A))^{op} \to \mathbb{A}$ is a sheaf with values in $\mathbb{A}$.

Proof : First, let us study the site $\mathcal{D}(\mathrm{Spec}_{\mathrm{Ind}}(A))$. Let T' be the Grothendieck topology on the category $\mathcal{D}(\mathrm{Spec}_{\mathrm{Ind}}(A))$ generated [21] by the empty family of morphisms of target $\emptyset$, and the families of morphisms of the form, $(u \to u \cup v, v \to u \cup v)$ satisfying $u \cap v = \emptyset$. As these families are covering for T, one has $T' \subset T$. We will show that $T' = T$. First of all, consider a family $(u \to w, v \to w)$ which is covering for T, i.e. such that $u \cup v = w$. Then, the family $(u \to w, v\text{-}u \to w)$ is covering for T'. The sieve R on w generated by $(u \to w, v \to w)$ contains the sieve R' on w generated by $(u \to w, v\text{-}u \to w)$. Since R' is covering for T', R is covering for T' and the family $(u \to w, v \to w)$ is covering for T'. Next, let us consider a non empty finite family $(u_i \to u)_{i\in[1,n]}$ which is a covering family for T, i.e., such that $\bigcup_{i=1}^{n} u_i = u$. This family is the composite of covering families for T, each with two members. According to what has already been shown, each of these two membered families is covering for T'. It follows that the composed family $(u_i \to u)_{i\in[1,n]}$ is covering for T'. Finally, let us consider a non empty family $(u_i \to u)_{i\in I}$ which is a covering family for T, i.e., such that $\bigcup_{i\in I} u_i = u$. The open sets of the topological base $\mathcal{D}(\mathrm{Spec}_{\mathrm{Ind}}(A))$, of the Stone space, $\mathrm{Spec}_{\mathrm{Ind}}(A)$, being compact, there exists a finite subset $I_0 \subset I$ such that $u = \bigcup_{i\in I_0} u_i$. The sieve R on u generated by the family $(u_i \to u)_{i\in I}$ contains the sieve R_0 on u generated by the subfamily $(u_i \to u)_{i\in I_0}$. Since the finite family $(u_i \to u)_{i\in I_0}$ is covering for T, it is covering for T', therefore R_0 is covering for T' as also are R and the family $(u_i \to u)_{i\in I}$. This completes the proof that $T' = T$, that is, that the topology T is generated by the empty family of morphisms of target $\emptyset$ and the families of the form $(u \to u \cup v, v \to u \cup v)$ where $u \cap v = \emptyset$. The set of families of this form is stable under inverse images. It follows, according to Cor. 2.3. and 2.4. Exp. II, of [21], that the functor $\tilde{A} : \mathcal{D}(\mathrm{Spec}_{\mathrm{Ind}}(A))^{op} \to \mathbb{A}$ is a sheaf with values in $\mathbb{A}$ if, and only if, $\tilde{A}(\emptyset) \simeq 1$ and, for each family $(u \to u \cup v,\ v \to u \cup v)$ satisfying $u \cap v = \emptyset$, the pair $(\tilde{A}(u \cup v) \to \tilde{A}(u),\ \tilde{A}(u \cup v) \to \tilde{A}(v))$ is a product. The condition $\tilde{A}(\emptyset) \simeq 1$ is satisfied, since $\tilde{A}(\emptyset) = \tilde{A}(D(0_A)) \simeq 1$. For the second condition, consider $\delta : A \to B$, $\mu : A \to C$ in $\Delta(A)$ satisfying $\delta \wedge \mu = 0_A$ and denote $\delta \vee \mu : A \to S$, $m : S \to B$, $n : S \to C$, the morphisms satisfying $m(\delta \vee \mu) = \delta$ and $n(\delta \vee \mu) = \mu$. Then $m = (\Delta(\delta \vee \mu))(\delta)$ and $n = (\Delta(\delta \vee \mu))(\mu)$. The two quotient objects m, $n \in \Delta(S)$ thus satisfy the relations $m \wedge n = (\Delta(\delta \vee \mu))(\delta \wedge \mu) = (\Delta(\delta \vee \mu))(0_A) = 0_S$ and $m \vee n = (\Delta(\delta \vee \mu))(\delta \vee \mu) = 1_S$ and, hence, they are complementary in $\Delta(S)$. This implies that the pair $(m : S \to B,\ n : S \to C)$ is a product (Proposition 1.3.3), i.e. the pair $(\tilde{A}(D(\delta \vee \mu)) \to \tilde{A}(D(\delta)),$ $\tilde{A}(D(\delta \vee \mu)) \to \tilde{A}(D(\mu)))$ is a product. ■

The sheaf $\tilde{A}$ defined on $\mathcal{D}(\mathrm{Spec}_{\mathrm{Ind}}(A))$ with values in $\mathbb{A}$ can be extended in a unique way to give a sheaf on $\mathrm{Spec}_{\mathrm{Ind}}(A)$ with values in $\mathbb{A}$, also denoted $\tilde{A}$ and called the structural sheaf on $\mathrm{Spec}_{\mathrm{Ind}}(A)$.

1.7.2. Proposition. The structural sheaf $\tilde{A}$ on $\mathrm{Spec}_{\mathrm{Ind}}(A)$ has, as stalks, the indecomposable components of A, and as its object of global sections, the object A.

Proof : At each point Φ of $\mathrm{Spec}_{\mathrm{Ind}}(A)$, the stalk of $\tilde{A}$ is $\varinjlim_{\delta \in \Phi} \tilde{A}(D(\delta)) = \varinjlim_{\delta \in \Phi} \beta(\delta) = A_\Phi$ (cf. proof 1.6.1). Moreover, $\tilde{A}(\mathrm{Spec}_{\mathrm{Ind}}(A)) = \tilde{A}(D(1_A)) = A$. ■

A Boolean space is a separated compact topological space having a topological base made up of sets which are both open and closed. For each Boolean space X, one denotes $\Omega(X)$, the category, i.e. the ordered set, of open sets of X and for any continuous function, $f : X \to Y$, one denotes $f^* : \Omega(Y)^{op} \to \Omega(X)^{op}$, the functor defined by $f^*u = f^{-1}(u)$. The category $\mathcal{S}hv\mathbb{B}ool\mathbb{I}nd\,\mathbb{A}$ of sheaves of indecomposable objects of $\mathbb{A}$ on Boolean spaces has as objects the pairs (X,F) where X is a Boolean space and $F : \Omega(X)^{op} \to \mathbb{A}$, a sheaf with base X and values in $\mathbb{A}$, the stalks F_x $(x \in X)$, being indecomposable objects of $\mathbb{A}$. It has as morphisms $(X,F) \to (Y,G)$ the pairs (f,α) where $f : Y \to X$ is a continuous map and $\alpha : F \to G\,f^*$ is a natural transformation. The global section functor $\Gamma : \mathbb{S}hv\mathbb{B}ool\mathbb{I}nd\ \mathbb{A} \to \mathbb{A}$ is defined by $\Gamma(X,F) = FX$ and $\Gamma(f,\alpha) = \alpha_X$. The structural sheaf functor $\Sigma : \mathbb{A} \to \mathbb{S}hv\mathbb{B}ool\mathbb{I}nd\ \mathbb{A}$ is defined by $\Sigma(A) = (\mathrm{Spec}_{\mathrm{Ind}}(A), \tilde{A})$ and $\Sigma(f) = (\mathrm{Spec}_{\mathrm{Ind}}(f), \sigma_f)$ where $(\sigma_f)_{D(\delta)}$ is the direct image of f by δ.

1.7.3. Theorem. Any locally indecomposable category $\mathbb{A}$ is equivalent to the category $\mathbb{S}hv\mathbb{B}ool\mathbb{I}nd\ \mathbb{A}$ of sheaves of indecomposable objects of $\mathbb{A}$ on Boolean spaces and the functors $\Sigma : \mathbb{A} \to \mathbb{S}hv\mathbb{B}ool\mathbb{I}nd\ \mathbb{A}$ and $\Gamma : \mathbb{S}hv\mathbb{B}ool\mathbb{I}nd\ \mathbb{A} \to \mathbb{A}$ are equivalences of categories, quasi-inverse to each other.

Proof : The functor $\Gamma\Sigma$ is the identity of $\mathbb{A}$, since $\Gamma\Sigma(A) = \Gamma(\tilde{A}) = \tilde{A}(\mathrm{Spec}_{\mathrm{Ind}}(A)) = A$ (Proposition 1.7.2), and, for any morphism $f : A \to B$ of $\mathbb{A}$, $\Gamma\Sigma(f) = \Gamma(\sigma_f) = (\sigma_f)_{D(1_A)} = f$. Let (X,F) be an object of $\mathbb{S}hv\mathbb{B}ool\mathbb{I}nd\ \mathbb{A}$. We will denote by $\mathcal{D}(X)$, the Boolean algebra of open-closed subsets of X. For $x \in X$, $u \in \mathcal{D}(X)$, $v \in \mathcal{D}(X)$ with $x \in u \cup v$, we will write $\rho^v_u : Fv \to Fu$, for the restriction morphism of F, F_x for the stalk of F at x and $\rho^u_x : Fu \to F_x$, the canonical injection. We denote the object FX by A. For each $u \in \mathcal{D}(X)$, the pair $(\rho^X_u : A \to Fu,\ \rho^X_{X-u} : A \to F(X-u))$ is a product, so the morphism $\rho^X_u : A \to Fu$ is a direct factor of A. This defines a function $\rho^X : \mathcal{D}(X) \to \Delta(A)$. This function is increasing and preserves the maximum element and the joins of those pairs of elements whose meet is the minimum element. It is therefore a homomorphism of Boolean algebras. It is injective, since $\rho^X_u = 0_A \Rightarrow Fu \simeq 1 \Rightarrow \prod_{x\in u} F_x \simeq 1 \Rightarrow u = \emptyset$, as the objects F_x, being indecomposable, are not isomorphic to 1. Let us show that it is surjective. Let $\delta : A \to B$ be an element of $\Delta(A)$. We will show that the set $u = \{x \in X : \rho^X_x$ can be factorised through $\delta\}$ is an open set of X. Let $x \in u$ and $m : B \to F_x$ be the unique morphism satisfying $m\delta = \rho^X_x$. Denote by $i_B : Z \to B$, $i_A : Z \to A$ and $i_v : Z \to Fv$, the uniquely defined morphisms and $f : Z^2 \to A$, the morphism such that (i_B, δ) is the amalgamated sum of (π, f) (Theorem 1.3.4). The two morphisms $i_A\pi$, $f : Z^2 \to A$ satisfy the relation $\rho^X_x i_A \pi = m i_B \pi = m\delta f = \rho^X_x f$. Since the object Z^2 is finitely presentable and $(\rho^v_x : Fv \to F_x)_{x\in v\in\mathcal{D}(X)}$ is a filtered colimit, there exists $w \in \mathcal{D}(X)$ such that $x \in w$ and $\rho^X_w i_A \pi = \rho^X_w f$, that is, such that $i_w\pi = \rho^X_w f$. Then, there exists a unique morphism $n : B \to Fw$, satisfying $n\delta = \rho^X_w$ and

$ni_B = i_w$. This implies that ρ_w^X factorises through δ and, consequently, that $x \in w \subset u$. This clearly shows that u is an open set of X. Similarly, if $\delta' : A \to B'$ is the complement of δ in $\Delta(A)$, the set $u' = \{x \in X : \rho_x^X$ factorises through $\delta'\}$ is an open set of X. Since the objects F_x are indecomposable, the sets u and u' are complementary in X ((4), proposition 1.5.1.) ; they are, therefore, closed in X and consequently belong to $\mathcal{D}(X)$. The morphism $(\rho_x^X)_{x\in u} : A \to \coprod_{x\in u} F_x$ factorises, both through the monomorphism $(\rho_x^u)_{x\in u} : Fu \to \coprod_{x\in u} F_x$ and through the morphism $\delta : A \to B$. However this latter is a regular epimorphism, as it is a direct image of the split epimorphism $\pi : Z^2 \to Z$. Therefore, there exists a morphism $r : B \to Fu$ satisfying $r\delta = \rho_u^X$, that is $\rho_u^X \leqslant \delta$. Similarly we can prove $\rho_{u'}^X \leqslant \delta'$. It follows that $\rho_u^X = \delta$, which ends the proof that the function $\rho^X : \mathcal{D}(X) \to \Delta(A)$ is an isomorphism of Boolean algebras. This isomorphism induces an isomorphism of the Stone space of $\mathcal{D}(X)$ on the Stone space of $\Delta(A)$ and, consequently, according to the Stone duality theorem, a homeomorphism of the space X onto $\mathrm{Spec}_{\mathrm{Ind}}(A)$. As the structural sheaf $\tilde{A}$ has as value, $\tilde{A}(D(\delta)) = \beta(\rho_u^X)$, one obtains an isomorphism between (X,F) and $(\mathrm{Spec}_{\mathrm{Ind}}(A),\tilde{A})$. It follows that the functor Σ is an equivalence of categories having as quasi-inverse the functor Γ. ∎

1.8. MORPHISMS OF LOCALLY INDECOMPOSABLE CATEGORIES.

One denotes by $U_o : \mathbb{B}\mathrm{ool} \to \mathbb{E}\mathrm{ns}$, the underlying set functor. If $\mathbb{A}$ and $\mathbb{B}$ are two locally indecomposable categories, one denotes $\Delta_{\mathbb{A}} : \mathbb{A} \to \mathbb{B}\mathrm{ool}$, and $\Delta_{\mathbb{B}} : \mathbb{B} \to \mathbb{B}\mathrm{ool}$, the functors, Δ, (1.3.5) associated respectively to $\mathbb{A}$ and $\mathbb{B}$. If $U : \mathbb{A} \to \mathbb{B}$ is a functor having a left adjoint, F, the image by F of the initial object Z of $\mathbb{B}$ is an initial object of $\mathbb{A}$ and, from now on, it will be assumed that $FZ = Z$. The <u>morphism</u> $e = (F\pi, F\pi') : F(Z^2) \to Z^2$ is then said to be <u>canonical</u>, similarly the <u>natural transformation</u>, $\varepsilon : \Delta_{\mathbb{A}} \to \Delta_{\mathbb{B}}U$, defined by $\varepsilon_A(\delta) = U\delta$, is said to be <u>canonical</u>. One says that a functor $U : \mathbb{A} \to \mathbb{B}$ <u>lifts direct factors</u> if, for each object A of $\mathbb{A}$ and each direct factor $\delta : UA \to B$ of UA, there is a direct factor $\mu : A \to M$ of A and an isomorphism $m : B \to UM$ satisfying $U\mu = m\delta$.

1.8.0. Definition. A morphism of locally indecomposable categories is a functor $U : \mathbb{A} \to \mathbb{B}$, whose source $\mathbb{A}$ and target $\mathbb{B}$ are two locally indecomposable categories, which preserves filtered colimits and has a left adjoint which preserves finite products.

1.8.1. Proposition. If $\mathbb{A}$ and $\mathbb{B}$ are two locally indecomposable categories and $U : \mathbb{A} \to \mathbb{B}$ is a functor which preserves filtered colimits and which has a left adjoint F, the following assertions are equivalent :

(1) U is a morphism of locally indecomposable categories,
(2) The canonical morphism $e : F(Z^2) \to Z^2$ is an isomorphism,
(3) The canonical morphism, $\varepsilon : \Delta_{\mathbb{A}} \to \Delta_{\mathbb{B}} U$ is an isomorphism,
(4) U reflects the final object 1 and lifts direct factors.

Proof : (1) => (2). The pair $(F\pi : F(Z^2) \to FZ,\ F\pi' : F(Z^2) \to FZ)$ is a product, thus the morphism $e = (F\pi, F\pi') : F(Z^2) \to Z^2$ is an isomorphism.

(2) => (1). The relations $U_o\Delta_{\mathbb{A}}(F1) \simeq \mathrm{Hom}_{\mathbb{A}}(Z^2,F1) \simeq \mathrm{Hom}_{\mathbb{A}}(F(Z^2),F1) \simeq \mathrm{Hom}_{\mathbb{B}}(Z^2,UF1) \simeq \mathrm{Hom}_{\mathbb{B}}(Z^2,1)$ imply that $\Delta_{\mathbb{A}}F(1) \simeq \{0\}$ and thus $F1 \simeq 1$. Let $(p : B\times C \to B,\ q : B\times C \to C)$ be a product in $\mathbb{B}$. Let $m : Z^2 \to B\times C$ be the morphism such that the direct image of π by m is p and the direct image of π' by m is q. Then the direct image of $F\pi$ (resp. $F\pi'$) by Fm is Fp (resp. Fq). As $(F\pi,F\pi')$ is a product, (Fp,Fq) is a product.

(2) <=> (3). The natural transformation $\mathrm{Hom}_{\mathbb{A}}(e,-) : \mathrm{Hom}_{\mathbb{A}}(Z^2,-) \to \mathrm{Hom}_{\mathbb{A}}(F(Z^2),-)$ is the composition of the isomorphism $\mathrm{Hom}_{\mathbb{A}}(Z^2,-) \to U_o\Delta_{\mathbb{A}}$, the natural transformation, $U_o\varepsilon : U_o\Delta_{\mathbb{A}} \to U_o\Delta_{\mathbb{B}}U$, the isomorphism $U_o\Delta_{\mathbb{B}}U \to \mathrm{Hom}_{\mathbb{B}}(Z^2,U(-))$ and the isomorphism $\mathrm{Hom}_{\mathbb{B}}(Z^2,U(-)) \to \mathrm{Hom}_{\mathbb{A}}(F(Z^2),-)$. Consequently, e is an isomorphism <=> $\mathrm{Hom}_{\mathbb{A}}(e,-)$ is an isomorphism <=> $U_o\varepsilon$ is an isomorphism <=> ε is an isomorphism.

(3) => (4). If A is an object of $\mathbb{A}$ such that $UA = 1$, then $\Delta_{\mathbb{A}}(A) = \Delta_{\mathbb{B}}UA = \{0\}$, thus $A \simeq 1$. If $\delta : UA \to B$ is a direct factor of UA, there is a direct factor $\mu : A \to M$ of A whose image by U is the direct factor δ, that is to say, there exists an isomorphism m satisfying $U\mu = m\delta$.

(4) => (3). For each object A of $\mathbb{A}$, the function $\varepsilon_A : \Delta_{\mathbb{A}}(A) \to \Delta_{\mathbb{B}}(UA)$ is a surjective homomorphism of Boolean algebras

satisfying : $\varepsilon_A(\delta) = 0_{UA} \Rightarrow \delta = 0$. It is thus an isomorphism. ■

1.8.2. Proposition. For any locally indecomposable category $\mathbb{A}$, the functor $\Delta_{\mathbb{A}} : \mathbb{A} \to \mathbb{B}ool$ is a morphism of locally indecomposable categories.

Proof : As the category $\mathbb{A}$ is cocomplete, the functor $U_o\Delta_{\mathbb{A}} \simeq Hom_{\mathbb{A}}(Z^2,-)$ has a left adjoint, F, such that $F(1) = Z^2$. Consequently the functor $\Delta_{\mathbb{A}}$ as a left adjoint K such that $K(Z^2) = Z^2$ [35, Theorem 21.5.3.]. The functor $\Delta_{\mathbb{A}}$ is thus a morphism of locally indecomposable categories (Proposition 1.8.1.). ■

1.8.3. Proposition. If $\mathbb{B}$ is a locally indecomposable category, a functor, $U : \mathbb{A} \to \mathbb{B}$, is a morphism of locally indecomposable categories if, and only if,

(1) $\mathbb{A}$ is locally finitely presentable,

(2) U preserves filtered colimits, has a left adjoint and lifts direct factors,

(3) the direct factor morphisms in $\mathbb{A}$ are strongly cocartesian for U, [19].

Proof : A morphism of locally indecomposable categories satisfies conditions (1) and (2) (definition 1.8.0 and Proposition 1.8.1). Let $\delta : A \to B$ be a direct factor in $\mathbb{A}$. If $g : A \to C$ and $m : UB \to UC$ are two morphisms satisfying $m(U\delta) = Ug$, the relation $(\Delta_{\mathbb{B}}(Ug))(U\delta) = 1_{UC}$ implies the relation $(\Delta_{\mathbb{A}}(g))(\delta) = 1_C$ ((3), Proposition 1.8.1) and thus the morphism g factorises via a morphism $n : B \to C$ satisfying $Un = m$. Direct factors in $\mathbb{A}$ are thus strongly cocartesian for U. Let us consider a functor $U : \mathbb{A} \to \mathbb{B}$ satisfying conditions (1), (2) and (3). Let $f : 1 \to A$ be a morphism in $\mathbb{A}$ and $t : A \to 1$, the unique morphism. As $U1$ is a strict final object of $\mathbb{B}$, the morphism Uf is an isomorphism and $(Uf)^{-1} = Ut$. The morphism $t : A \to 1$, being a direct factor, is strongly cocartesian, thus there exists a morphism $s : 1 \to A$ satisfying $Us = Uf$ and $st = 1_A$. It follows that t is an isomorphism as is f. The final object of $\mathbb{A}$ is thus strict. Let $(p : A\times B \to A,$ $q : A\times B \to B)$ be a product in $\mathbb{A}$. If $(m : A \to C,$ $n : B \to C)$ is the amalgamated sum of (p,q), the image UC is necessarily the final object thus C is the final object by a proof, analogous to the one just used, using the direct factor $C \to 1$. Finite products are thus codisjoint in $\mathbb{A}$. Let $f : A\times B \to D$ and let $(r : UA \to R,$ $s : UD \to R)$ denote the amalgamated sum of (Up,Uf). As p is a direct factor, the morphisms Up and s are direct factors. There is a direct factor $n : D \to E$ and an isomorphism $t : R \to UE$ satisfying $ts = Un$. The morphism p being strongly cocartesian for U, there is a morphism $m : A \to E$ satisfying $mp = nf$ and $Um = tr$. The morphism n being strongly co-cartesian for U, the pair (n,m) is the amalgamated sum of (f,p). Similarly, one constructs the amalgamated sum $(n_1 : D \to E_1, \; m_1 : B \to E_1)$ of (f,q). The pair (Un, Un_1) is then a product in $\mathbb{B}$. Let $n' : D \to E'$ a morphism which forms, with n, a product (n,n'). Then (Un,Un') is a product. The direct factors Un_1 and Un' are both complements of the direct factor Un. There is thus an isomorphism $r : UE_1 \to UE'$ satisfying $r(Un_1) = Un'$. As n_1 and n' are morphisms which are strongly cocartesian for U, there is a morphism $v : E_1 \to E'$ such that $vn_1 = n'$ and it is an isomorphism. The pair (n,n_1) is thus a product. Finite products are thus couniversal in $\mathbb{A}$. It remains to prove that the object Z^2 is finitely presentable in $\mathbb{A}$. First let us note that the existence of the functor $\Delta_{\mathbb{A}} : \mathbb{A} \to \mathbb{B}ool$ and of the isomorphism $U_o\Delta_{\mathbb{A}} \simeq Hom_{\mathbb{A}}(Z^2,-)$ do not depend on the fact that Z^2 is finitely presentable (cf. 1.3.5.). Then we note the isomorphism : $\Delta_{\mathbb{A}} \simeq \Delta_{\mathbb{B}}U$. The functor $Hom_{\mathbb{A}}(Z^2,-) \simeq U_o\Delta_{\mathbb{A}} \simeq U_o\Delta_{\mathbb{B}}U$, thus preserves filtered colimits. This implies that $\mathbb{A}$ is locally indecomposable and that U is a morphism of locally indecomposable categories (Proposition 1.8.1). ∎

1.9. FULL LOCALLY INDECOMPOSABLE SUBCATEGORIES.

A locally indecomposable subcategory of a locally indecomposable category $\mathbb{B}$ is a subcategory $\mathbb{A}$ of $\mathbb{B}$ such that the inclusion functor $\mathbb{A} \to \mathbb{B}$ is a morphism of locally indecomposable categories.

A subcategory $\mathbb{A}$ of $\mathbb{B}$ is said to be closed for direct factors if every direct factor in $\mathbb{B}$, whose source is in $\mathbb{A}$, is a morphism in $\mathbb{A}$.

1.9.0. Proposition. If $\mathbb{B}$ is a locally indecomposable category and $\mathbb{A}$ is a full reflexive subcategory of $\mathbb{B}$, closed under filtered colimits, the following assertions are equivalent :

(1) $\mathbb{A}$ is a locally indecomposable subcategory of $\mathbb{B}$,

(2) $\mathbb{A}$ is closed for direct factors,

(3) an object of $\mathbb{B}$ belongs to $\mathbb{A}$ if, and only if, its indecomposable components belong to $\mathbb{A}$.

Proof : The equivalence (1) <=> (2) results from Proposition 1.8.3.

(3) => (2). Let A be an object of $\mathbb{A}$ and $\delta : A \to B$ a direct factor of A. Any indecomposable component $\delta_\Phi : B \to B_\Phi$ of B is such that $\delta_\Phi \delta : A \to B$ is an indecomposable component of A and thus belongs to $\mathbb{A}$. It follows that B belongs to $\mathbb{A}$.

(2) => (3). If A is an object of $\mathbb{A}$, its indecomposable components belong to $\mathbb{A}$ as they are filtered colimits of direct factors of A. Let B be an object of $\mathbb{B}$ all of whose indecomposable components belong to $\mathbb{A}$. Let us denote by X, the topological space $\mathrm{Spec}_{\mathrm{Ind}}(B)$ and by $|X|$, the underlying set. Let $\mathcal{S}hv(X,\mathbb{B})$ be the category of sheaves on X with values in $\mathbb{B}$ and $\mathbb{B}^{|X|}$, the category of families of objects in $\mathbb{B}$ indexed by $|X|$. Following D.H. Van Osdol [37], let us denote $S : \mathcal{S}hv(X,\mathbb{B}) \to B^{|X|}$ the stalk functor and $Q : \mathbb{B}^{|X|} \to \mathcal{S}hv(X,\mathbb{B})$, the functor defined by $Q((B_x)_{x\in X})(V) = \prod_{x\in V} B_x$. By Theorem III (3) [37], when the category $\mathbb{B}$ is a category of algebraic structures, the functor S is left adjoint to the functor Q and is comonadic. But one sees easily that this theorem is also true for any finitely presentable category $\mathbb{B}$. As a result one has that the sheaf $\tilde{B} \in \mathcal{S}hv(X,\mathbb{B})$ has a canonical presentation as an equaliser of two morphisms of source $QS(\tilde{B})$ and of target $QSQS(\tilde{B})$, [33, p.149], in the category $\mathcal{S}hv(X,\mathbb{B})$, and hence that the object $B = \tilde{B}(X)$ has a canonical presentation as an equaliser of two morphisms of source $\prod_{\Phi\in X} B_\Phi$ and target $\prod_{\psi\in X} \varinjlim_{\psi\in V}(\prod_{\Phi\in V} B_\Phi)$, where the B_Φ are the indecomposable components of B (Proposition 1.6.1) thus belonging to $\mathbb{A}$. As $\mathbb{A}$ is closed in $\mathbb{B}$ for products, cofiltered limits and kernels , it results that B belongs to $\mathbb{A}$. ■

Let us recall that a variety of a category of algebras $\mathbb{A}lg(\mathbb{T})$ is a full subcategory of $\mathbb{A}lg(\mathbb{T})$ having for its objects the $\mathbb{T}$-algebras which satisfy a set of identities of $\mathbb{T}$ [35, Theorem 21.7.9.].

1.9.1. Proposition. Any variety of a locally indecomposable category of algebras, $\mathbb{A}lg(\mathbb{T})$, is a full locally indecomposable subcategory.

Proof : Any variety $\mathbb{K}$ of $\mathbb{A}lg(\mathbb{T})$ is a full reflexive subcategory, closed under products, subobjects, quotient algebras and filtered colimits [35, Theorem 21.7.0]. The result follows then from Proposition 1.9.0. ■

1.9.2. Examples. The following categories (cf. 1.1.2) are locally indecomposable being full locally indecomposable subcategories of the category $\mathbb{R}$ngc : $\mathbb{R}$ngc(p), p-$\mathbb{R}$ngc, p^k-$\mathbb{R}$ngc, $\mathbb{R}$ngc$\mathbb{R}$ed, $\mathbb{R}$ngc$\mathbb{R}$eg, $\mathbb{R}$ngc$\mathbb{F}$orm$\mathbb{R}$l, $\mathbb{R}$ncg$\mathbb{O}$, $\mathbb{R}$ngc$\mathbb{R}$ed$\mathbb{O}$, $\mathbb{R}$ngc$\mathbb{O}$$\mathbb{A}nn\mathbb{C}$onv, $\mathbb{R}$ngc$\mathbb{R}$ep$\mathbb{D}$om, $\mathbb{B}$ool. The categories : $\mathbb{R}$ng$\mathbb{R}$ed, $\mathbb{R}$ng$\mathbb{S}$t$\mathbb{R}$eg are full locally indecomposable subcategories of the category $\mathbb{R}$ng$\mathbb{S}$ym. The categories $\mathbb{R}$ngc$\mathbb{D}$if(p), $\mathbb{R}$ngc$\mathbb{D}$if$\mathbb{R}$ed, $\mathbb{R}$ngc$\mathbb{D}$if$\mathbb{R}$eg, $\mathbb{R}$ngc$\mathbb{D}$if$\mathbb{R}$eg$\mathbb{P}$er(p) are full locally indecomposable subcategories of $\mathbb{R}$ngc$\mathbb{D}$if. The categories $\mathbb{R}$ngc$\mathbb{O}$rd$\mathbb{R}$ed, $\mathbb{R}$ng$\mathbb{O}$rd$\mathbb{A}$nn$\mathbb{C}$onv are full locally indecomposable categories of $\mathbb{R}$ngc$\mathbb{O}$rd. The categories $\mathbb{R}$ngc$\mathbb{S}$t$\mathbb{L}$at, $\mathbb{R}$ngc$\mathbb{S}$t$\mathbb{L}$at $\mathbb{P}$roj, $\mathbb{R}$ncg$\mathbb{S}$t$\mathbb{L}$at$\mathbb{R}$ep$\mathbb{D}$om are full locally indecomposable subcategories of $\mathbb{R}$ncg$\mathbb{L}$at.

1.10. LOCALLY INDECOMPOSABLE CATEGORIES OF FRACTIONS.

Let $\mathbb{A}$ be a locally indecomposable category and Σ a set of finitely presentable morphisms of $\mathbb{A}$, i.e. whose source and target are finitely presentable objects. We will show that there is a universal morphism of locally indecomposable categories $U : \mathbb{A}' \to \mathbb{A}$, whose left adjoint F sends the morphisms of Σ onto isomorphisms ; the category $\mathbb{A}'$ is then the locally indecomposable category of fractions of $\mathbb{A}$ defined by Σ.

Following P. Gabriel and M. Zisman [17, definition 4.1], an object A of $\mathbb{A}$ is said to be left closed for Σ if the functions $\mathrm{Hom}_{\mathbb{A}}(\sigma, A)$ are bijective for all $\sigma \in \Sigma$. The full subcategory of $\mathbb{A}$, whose objects are the objects of $\mathbb{A}$ which are left closed for Σ, will be denoted $\mathbb{A}_\Sigma$.

1.10.0. Definition. An object A of $\mathbb{A}$ is locally left closed for Σ if each indecomposable component of A is left closed for Σ.

The full subcategory of $\mathbb{A}$, whose objects are the objects of $\mathbb{A}$ which are locally left closed for Σ, will be denoted $\mathbb{A}_{\mathrm{loc}\,\Sigma}$.

1.10.1. Theorem. The category $\mathbb{A}_{\mathrm{loc}\,\Sigma}$ of objects of $\mathbb{A}$ which are locally left closed for Σ, is a full locally indecomposable subcategory of $\mathbb{A}$, whose reflector sends the morphisms of Σ onto isomorphisms, and it is such that any morphism of locally indecomposable categories, $U : \mathbb{B} \to \mathbb{A}$, whose left adjoint sends the morphisms of Σ

<u>onto isomorphisms, factorises uniquely via</u> $\mathbb{A}_{loc\Sigma}$ <u>by a morphism of locally indecomposable categories</u>.

<u>Proof</u> : We will denote by $\bar{\Sigma}$, the set of morphisms of the form $\sigma\times 1_Z : X\times Z \to Y\times Z$ with $\sigma : X \to Y$ in Σ. The morphisms in $\bar{\Sigma}$ are finitely presentable (Proposition 1.4.0.), and thus the category $\mathbb{A}_{\bar{\Sigma}}$ is a full reflexive subcategory of $\mathbb{A}$ closed under filtered colimits [16]. Let A, B be two objects of $\mathbb{A}$ such that $A\times B \in \mathbb{A}_{\bar{\Sigma}}$. For any morphism $\sigma : X \to Y$ in Σ, the two functions $Hom_{\mathbb{A}}(\sigma\times 1_Z, A)$ and $Hom_{\mathbb{A}}(\sigma\times 1_Z, B)$ have non empty sets as sources and have as their product the bijection $Hom_{\mathbb{A}}(\sigma\times 1_Z,A)\times Hom_{\mathbb{A}}(\sigma\times 1_Z,B) \to Hom_{\mathbb{A}}(\sigma\times 1_Z, A\times B)$; they are thus themselves bijections. It follows that $A \in \mathbb{A}_{\bar{\Sigma}}$ and $B \in \mathbb{A}_{\bar{\Sigma}}$. The category $\mathbb{A}_{\bar{\Sigma}}$ is thus a full locally indecomposable subcategory of $\mathbb{A}$ (Proposition 1.9.0). For an indecomposable object A of $\mathbb{A}$ and a morphism $\sigma \in \Sigma$, the function $Hom_{\mathbb{A}}(\sigma\times 1_Z,A)$ is the direct sum of the two functions $Hom_{\mathbb{A}}(\sigma,A)$ and $Hom_{\mathbb{A}}(1_Z,A)$ (Proposition 1.5.1), consequently : $A \in \mathbb{A}_{\bar{\Sigma}} \iff Hom_{\mathbb{A}}(\sigma\times 1_Z,A)$ is a bijection for every $\sigma \in \Sigma \iff A \in \mathbb{A}_{\Sigma}$. For any object A of $\mathbb{A}$, one thus has : $A \in \mathbb{A}_{\bar{\Sigma}} \iff$ every indecomposable component of A belongs to $\mathbb{A}_{\bar{\Sigma}} \iff$ every indecomposable component of A belongs to $\mathbb{A}_{\Sigma} \iff A \in \mathbb{A}_{loc\Sigma}$. Thus $\mathbb{A}_{loc\Sigma} = \mathbb{A}_{\bar{\Sigma}}$. We will denote by $R : \mathbb{A} \to \mathbb{A}_{loc\Sigma}$, the reflector. For $\sigma \in \Sigma$ and any object $A \in \mathbb{A}_{loc\Sigma}$, the morphism $R(\sigma\times 1_Z) \simeq R\sigma\times R(1_Z)$ is an isomorphism and so $R\sigma$ must be an isomorphism. Let $U : \mathbb{B} \to \mathbb{A}$ be a morphism of locally indecomposable categories whose left adjoint, F, sends the the morphisms in Σ to isomorphisms. For each object B of $\mathbb{B}$, and each $\sigma \in \Sigma$, the function $Hom_{\mathbb{A}}(\sigma\times 1_Z,UB) \simeq Hom_{\mathbb{B}}(F(\sigma\times 1_Z),B) \simeq Hom_{\mathbb{B}}(F\sigma\times F(1_Z),B)$ is a bijection, so $UB \in \mathbb{A}_{\bar{\Sigma}} = \mathbb{A}_{loc\Sigma}$. The functor U thus factorises through the inclusion functor $J : \mathbb{A}_{loc\Sigma} \to \mathbb{A}$ and a functor $V : \mathbb{B} \to \mathbb{A}_{loc\Sigma}$ which is a morphism of locally indecomposable categories, since V preserves filtered colimits and has as its left adjoint the functor FJ, which preserves finite products. ■

An object X of $\mathbb{A}$ is said to be <u>weakly initial</u> [33, p.231] if for any object A of $\mathbb{A}$, there is at least one morphism from X to A, or what amounts to the same thing here, if there is at least one morphism from X to Z.

1.10.2. <u>Proposition</u>. <u>If Σ is a set of morphisms in $\mathbb{A}$, whose sources are weakly initial, then the objects of $\mathbb{A}$ which are locally left closed for Σ are exactly the objects which are left closed for Σ, i.e.</u> $\mathbb{A}_{loc\Sigma} = \mathbb{A}_{\Sigma}$.

<u>Proof</u> : Let A and B be two objects of $\mathbb{A}$ and $\sigma : X \to Y$ be a morphism in Σ. As the sets $\mathrm{Hom}_{\mathbb{A}}(X,A)$ and $\mathrm{Hom}_{\mathbb{A}}(X,B)$ are non empty, the function $\mathrm{Hom}_{\mathbb{A}}(\sigma,A\times B) \simeq \mathrm{Hom}_{\mathbb{A}}(\sigma,A)\times\mathrm{Hom}_{\mathbb{A}}(\sigma,B)$ is a bijection if, and only if, both of the functions $\mathrm{Hom}_{\mathbb{A}}(\sigma,A)$ and $\mathrm{Hom}_{\mathbb{A}}(\sigma,B)$ are. It follows that $A\times B \in \mathbb{A}_{\Sigma} \iff A \in \mathbb{A}_{\Sigma}$ and $B \in \mathbb{A}_{\Sigma}$. As a result, $\mathbb{A}$ is a full locally indecomposable subcategory of $\mathbb{A}$ and, for any object A of $\mathbb{A}$, one has : $A \in \mathbb{A}_{\Sigma} \iff$ the indecomposable components of A belong to $\mathbb{A}_{\Sigma} \iff A \in \mathbb{A}_{loc\Sigma}$. ■

Let Γ be a set of finite families of morphisms with a common source and whose sources and targets are finitely presentable. Following [13, 3.17], an object A of $\mathbb{A}$ is <u>strictly left closed for</u> Γ if, for any family $(\gamma_i : X \to X_i)_{i\in[1,n]}$ in Γ, any morphism $f : X \to A$ factorises in a unique way via a unique morphism γ_i.

1.10.3. <u>Definition</u>. An object A of $\mathbb{A}$ is <u>locally strictly left closed for</u> Γ if the indecomposable components of A are strictly left closed for Γ.

These objects generate the full subcategory $\mathbb{A}_{loc(\Gamma)}$ of $\mathbb{A}$.

1.10.4. <u>Theorem</u>. <u>The category</u> $\mathbb{A}_{loc(\Gamma)}$ <u>of objects of $\mathbb{A}$ which are locally strictly left closed for</u> Γ <u>is a full locally indecomposable subcategory of</u> $\mathbb{A}$ <u>whose reflector sends the families of morphisms of</u> Γ <u>onto products and is such that any morphism of locally indecomposable categories,</u> $U : \mathbb{B} \to \mathbb{A}$, <u>whose left adjoint sends the families of morphisms from</u> Γ <u>onto products, factorises in a unique way via</u> $\mathbb{A}_{loc(\Gamma)}$ <u>by a morphism of locally indecomposable categories</u>.

<u>Proof</u> : For each family $(\gamma_i : X \to X_i)_{i\in[1,n]}$ in Γ, let us denote by $(p_i : \prod_{i=1}^{n} X_i \to X_i)_{i\in[1,n]}$, the product of the X_i and by $\gamma : X \to \prod_{i=1}^{n} X_i$, the morphism defined by $p_i\gamma = \gamma_i$ for every $i\in[1,n]$. The morphisms, γ, are finitely presentable (Proposition 1.4.0). For each indecomposable object A of $\mathbb{A}$, the function

$\langle\mathrm{Hom}_{\mathbb{A}}(p_i,A)\rangle_{i\in[1,n]} : \coprod_{i=1}^{n} \mathrm{Hom}_{\mathbb{A}}(X_i,A) \to \mathrm{Hom}_{\mathbb{A}}(\prod_{i=1}^{n} X_i,A)$ is a bijection (Proposition 1.5.1.), and the function, $\langle \mathrm{Hom}_{\mathbb{A}}(\gamma_i, A)\rangle_{i\in[1,n]} : \coprod_{i=1}^{n} \mathrm{Hom}_{\mathbb{A}}(X_i, A) \to \mathrm{Hom}_{\mathbb{A}}(X, A)$, satisfies $\mathrm{Hom}_{\mathbb{A}}(\gamma, A) \circ \langle \mathrm{Hom}_{\mathbb{A}}(p_i, A)\rangle_{i\in[1,n]} = \langle \mathrm{Hom}_{\mathbb{A}}(\gamma_i, A)\rangle_{i\in[1,n]}$. It follows that A is left closed for γ if, and only if, A is strictly left closed for $(\gamma_i : X \to X_i)_{i\in[1,n]}$. If one denotes by Σ, the set of morphisms of the form γ for $(\gamma_i : X \to X_i)_{i\in[1,n]}$ an element of Γ, then $\mathbb{A}_{\mathrm{loc}(\Gamma)} = \mathbb{A}_{\mathrm{loc}\Sigma}$ is a full locally indecomposable subcategory of $\mathbb{A}$ (Theorem 1.10.1). Moreover, for any morphism of locally indecomposable categories, $U : \mathbb{B} \to \mathbb{A}$, the left adjoint, F, sends the morphism $\gamma \in \Sigma$ to an isomorphism if, and only if, it sends the family $(\gamma_i : X \to X_i)_{i\in[1,n]}$ to a product. The result thus follows from Theorem 1.10.1.

∎

1.10.5. Examples. The category $\mathbb{R}\mathrm{ngc}(p)$ is the category $\mathbb{R}\mathrm{ngc}_{\{\sigma_1\}}$, where $\sigma_1 : \mathbb{Z} \to \mathbb{Z}/p\mathbb{Z}$ is the quotient morphism. The category p-$\mathbb{R}\mathrm{ngc}$ is the category $\mathbb{R}\mathrm{ngc}_{\{\sigma_1,\sigma_2\}}$ where $\sigma_2 : \mathbb{Z}[X] \to \mathbb{Z}[X]/(X^p - X)$ is the quotient morphism. The category p^k-$\mathbb{R}\mathrm{ngc}$ is the category $\mathbb{R}\mathrm{ngc}_{\{\sigma_1,\sigma_3\}}$ where $\sigma_3 : \mathbb{Z}[X] \to \mathbb{Z}[X]/(X^{p^k} - X)$ is the quotient morphism. The category $\mathbb{R}\mathrm{ngc}\mathbb{R}\mathrm{ed}$, is the category $\mathbb{R}\mathrm{ngc}_{\{\sigma_4\}}$ where $\sigma_4 : \mathbb{Z}[X]/(X^2) \to \mathbb{Z}$ is the morphism which kills X. The category $\mathbb{R}\mathrm{ngc}\mathbb{R}\mathrm{eg}$ is the category $\mathbb{R}\mathrm{ngc}_{\{\sigma_5\}}$ where $\sigma_5 : \mathbb{Z}[X] \to \mathbb{Z}[X,Y]/(X^2Y - X) + (Y^2X - Y)$ is defined by $\sigma_5(X) = \bar{X}$.

1.11. LOCALLY INDECOMPOSABLE CATEGORIES OF ALGEBRAS.

If $\mathbb{T} = (T, \eta, \mu)$ is a monad on a category $\mathbb{A}$, $\mathbb{A}^{\mathbb{T}}$ denotes the category of $\mathbb{T}$-algebras and $U^{\mathbb{T}} : \mathbb{A}^{\mathbb{T}} \to \mathbb{A}$ the forgetful functor [33, p.136]. The monad $\mathbb{T}$ is said to be of finite rank [16] if the functor, T, preserves filtered colimits. $\mathbb{T}$ is said to preserve finite products if the functor T preserves them.

1.11.0. Proposition. *If $\mathbb{A}$ is a locally indecomposable category and $\mathbb{T}$ is a monad on $\mathbb{A}$, of finite rank, which preserves finite products, then the forgetful functor $U^{\mathbb{T}} : \mathbb{A}^{\mathbb{T}} \to \mathbb{A}$ is a morphism of locally indecomposable categories.*

Proof : Let us show that the functor $U^{\mathbb{T}} : \mathbb{A}^{\mathbb{T}} \to \mathbb{A}$ satisfies the conditions of Proposition 1.8.3. The category $\mathbb{A}^{\mathbb{T}}$ is locally finitely presentable and the functor $U^{\mathbb{T}}$ preserves filtered colimits and has a left adjoint [16]. Let $(p : A\times B \to A,\ q : A\times B \to B)$ be a product and $(A\times B, c)$ a $\mathbb{T}$-algebra. The pair $(Tp : T(A\times B) \to TA,\ Tq : T(A\times B) \to TB)$ being a product, the relations $Tp \leqslant \Delta(\eta_{A\times B})(p)$ and $Tq \leqslant \Delta(\eta_{A\times B})(q)$ imply $Tp = \Delta(\eta_{A\times B})(p)$ and $Tq = \Delta(\eta_{A\times B})(q)$. Thus (Tp, η_A) is an amalgamated sum of $(\eta_{A\times B}, p)$. As the morphisms $pc = T(A\times B) \to A$ and $1_A : A \to A$ satisfy $pc\eta_{A\times B} = 1_A p$, there exists a morphism $a : TA \to A$ satisfying $a\eta_A = 1_A$ and $a(Tp) = pc$. The relations, $a(Ta)(T^2p) = a(T(a(Tp))) = aT(pc) = a(Tp)(Tc) = pc(Tc) = pc\mu_{A\times B} = a(Tp)\mu_{A\times B} = a\mu_A(T^2p)$, together with the fact that T^2p is an epimorphism imply the equality $a(Ta) = a\mu_A$. The pair (A,a) is thus a $\mathbb{T}$-algebra and $p : (A\times B, c) \to (A,a)$ is a morphism of $\mathbb{T}$-algebras. Similarly, one defines a $\mathbb{T}$-algebra (B,b) with a morphism $q : (A\times B, c) \to (B,b)$. The pair (p,q) is then a product in $\mathbb{A}^{\mathbb{T}}$. In turn, this proves that $p : (A\times B, c) \to (B,b)$ is a direct factor in $\mathbb{A}^{\mathbb{T}}$. The functor $U^{\mathbb{T}}$ thus lifts direct factors. Let us show that direct factors in $\mathbb{A}^{\mathbb{T}}$ are strongly cocartesian for $U^{\mathbb{T}}$. Let $\delta : (A,a) \to (D,d)$ be one of them. The morphism δ and $T\delta$ are direct factors in $\mathbb{A}$ and are thus epimorphisms. Let $g : (A,a) \to (E,e)$ be a morphism of $\mathbb{T}$-algebras and $m : D \to E$ a morphism of $\mathbb{A}$ satisfying $m\delta = g$. The relations $e(Tm)(T\delta) = eT(m\delta) = e(Tg) = ga = m\delta a = md(T\delta)$, together with the fact that $T\delta$ is an epimorphism, imply the equality $e(Tm) = md$. The morphism m is thus a morphism of $\mathbb{T}$-algebras $(D,d) \to (E,e)$, from which it follows that $\delta : (A,a) \to (D,d)$ is strongly cocartesian for $U^{\mathbb{T}}$. ■

1.11.1. <u>Proposition</u>. <u>If $\mathbb{T}$, $\mathbb{T}'$ are two algebraic theories in the sense of F. W. Lawvere, an algebraic functor $U : \mathbb{A}lg(\mathbb{T}') \to \mathbb{A}lg(\mathbb{T})$ between the corresponding categories of algebras is a morphism of locally indecomposable categories if, and only if, the category $\mathbb{A}lg(\mathbb{T})$ is locally indecomposable and U lifts direct factors.</u>

<u>Proof</u> : The result follows from [35, Theorem 18.5.3] and from Proposition 1.8.3 since the direct factors of $\mathbb{A}lg(\mathbb{T}')$ are the surjective $\mathbb{T}'$-morphisms and that these are strongly cocartesian for U. ■

1.11.2. <u>Examples</u>. The categories $\mathbb{R}ngc\mathbb{B}a$, $\mathbb{R}ngc\mathbb{D}if$, $\mathbb{R}ngc\mathbb{L}at$, $\mathbb{R}ngc\mathbb{S}t\mathbb{L}at\mathbb{B}a$ are categories algebraic over the category $\mathbb{R}ngc$ and it is immediate that the functors forgetting structure, with values in $\mathbb{R}ngc$, lift direct factors. As we know that the category $\mathbb{R}ngc$ is locally indecomposable (1.2.1), one can deduce that these categories are also locally indecomposable. Similarly, knowing that the category $\mathbb{L}at\mathbb{D}ist$ is locally indecomposable, one deduces that this is also true for the category $\mathbb{H}eyt$.

If A is an object of a category, $\mathbb{A}$, <u>the category of objects of $\mathbb{A}$ under A</u> [33, p. 46] is the category $A/\mathbb{A}$, whose objects are the pairs (X,f) where X is an object of $\mathbb{A}$ and $f : A \to X$, a morphism, and in which the morphisms $(X,f) \to (Y,g)$ are the morphisms $h : X \to Y$ of $\mathbb{A}$ satisfying $hf = g$. The target functor $F : A/\mathbb{A} \to \mathbb{A}$ is defined by $F(X,f) = X$ and $F(h) = h$.

1.11.3. <u>Proposition</u>. <u>If $\mathbb{A}$ is a locally indecomposable category and A is an object of $\mathbb{A}$, the target functor $A/\mathbb{A} \to \mathbb{A}$ is a morphism of locally indecomposable categories</u>.

<u>Proof</u> : The result follows from Proposition 1.11.0., because the target functor $A/\mathbb{A} \to \mathbb{A}$ is monadic of finite rank and has a left adjoint, the functor $A \coprod (-)$, which preserves finite products. ■

1.11.4. <u>Examples</u>. If $A \in \mathbb{R}ngc$, the category $A/\mathbb{R}ngc$ is isomorphic to the category $\mathbb{A}lg(A)$ of algebras over A. Starting from different locally indecomposable categories of rings, one can, in this way, construct locally indecomposable categories of algebras.

1.12. <u>LOCALLY INDECOMPOSABLE CATEGORIES OF COALGEBRAS FOR A COMONAD</u>.

We will use the notion of comonad, dual that of monad [33].

1.12.0. <u>Definition</u>. A <u>comonad</u> $\mathbb{L} = (L,\varepsilon,\delta)$ on a category $\mathbb{A}$ is said to be <u>of finite rank</u> if the functor L preserves filtered colimits and if any morphism $f : X \to A$, from a finitely presentable object X of $\mathbb{A}$ to an object A, which underlies an $\mathbb{L}$-coalgebra (A,a), factorises as $f = hg$ where $g : X \to B$ is a morphism whose target is a finitely presentable object of $\mathbb{A}$ and $h : B \to A$ is a morphism underlying a $\mathbb{L}$-coalgebras morphism $h : (B,b) \to (A,a)$.

1.12.1. <u>Proposition</u>. <u>If $\mathbb{A}$ is a locally finitely presentable category and $\mathbb{L}$ is a comonad of finite rank on $\mathbb{A}$, the category $\mathbb{A}^{\mathbb{L}}$ of $\mathbb{L}$-coalgebras is locally finitely presentable, and its finitely presentable objects are precisely the $\mathbb{L}$-coalgebras on the finitely presentable</u>

objects of $\mathbb{A}$.

Proof : Let $\mathbb{L} = (L,\varepsilon,\delta)$, $U^{\mathbb{L}} : \mathbb{A}^{\mathbb{L}} \to \mathbb{A}$ be the forgetful functor and $D^{\mathbb{L}} : \mathbb{A} \to \mathbb{A}^{\mathbb{L}}$ its right adjoint. The functor $U^{\mathbb{L}}$ creates colimits, so the category $\mathbb{A}^{\mathbb{L}}$ is cocomplete and the colimits in $\mathbb{A}^{\mathbb{L}}$ are preserved by $U^{\mathbb{L}}$. As L preserves filtered colimits, $D^{\mathbb{L}}$ preserves them also. If (A,a) is a finitely presentable $\mathbb{L}$-coalgebra, the functor $\mathrm{Hom}_{\mathbb{A}}(A,-) = \mathrm{Hom}_{\mathbb{A}}(U^{\mathbb{L}}(A,a),-) \simeq \mathrm{Hom}_{\mathbb{A}^{\mathbb{L}}}((A,a),D^{\mathbb{L}}(-))$ preserves filtered colimits and consequently A is a finitely presentable object of $\mathbb{A}$. Conversely, consider an $\mathbb{L}$-coalgebra, (A,a), on a finitely presentable object A of $\mathbb{A}$. Let $(\alpha_i : (X_i,x_i) \to (X,x))_{i\in\mathbb{I}}$ be a filtered colimit in $\mathbb{A}^{\mathbb{L}}$. The cones $(\alpha_i : X_i \to X)_{i\in\mathbb{I}}$ and $(L\alpha_i : LX_i \to LX)_{i\in\mathbb{I}}$ are filtered colimits in $\mathbb{A}$. Let $f : (A,a) \to (X,x)$. There is an index $i \in \mathbb{I}$ and a morphism $f_i : A \to X_i$ such that $f = \alpha_i f_i$. The relations $(L\alpha_i)x_i f_i = x\alpha_i f_i = xf = (Lf)a = L(\alpha_i f_i)a = (L\alpha_i)(Lf_i)a$ imply the existence of a morphism $u : i \to j$ of $\mathbb{I}$ satisfying $(LX_u)x_i f_i = (LX_u)(Lf_i)a$, so $x_j X_u f_i = L(X_u f_i)a$. One thus obtains a morphism $X_u f_i : (A,a) \to (X_j,x_j)$, which factorises the morphism $f : (A,a) \to (X,x)$ in the form $f = \alpha_j(X_u f_i)$. Moreover, if $f_i : (A,a) \to (X_i,x_i)$ and $f_j : (A,a) \to (X_j,x_j)$ are two morphisms satisfying $\alpha_i f_i = \alpha_j f_j$, there are two morphisms $u : i \to k$ and $v : j \to k$ in $\mathbb{I}$ such that $X_u f_i = X_v f_j$. It follows that the functor $\mathrm{Hom}_{\mathbb{A}^{\mathbb{L}}}((A,a),-)$ preserves filtered colimits, that is to say, that the object (A,a) is finitely presentable in $\mathbb{A}^{\mathbb{L}}$. The functor $U^{\mathbb{L}}$ thus preserves and reflects the finitely presentable objects. Let (A,a) be any object of $A^{\mathbb{L}}$. In the category $\mathbb{A}$, the object A is a filtered colimit of finitely presentable objects : say, $(A,(\alpha_i)) = \varinjlim_{i\in\mathbb{I}} A_i$. As the comonad $\mathbb{L}$ is of finite rank, each morphism $\alpha_i : A_i \to A$ factorises in the form $\alpha_i = \gamma_i\beta_i$, where $\beta_i : A_i \to B_i$ is a morphism of $\mathbb{A}$ whose target is finitely presentable and $\gamma_i : (B_i,b_i) \to (A,a)$ is a morphism of $\mathbb{L}$-coalgebras. The $\mathbb{L}$-coalgebras (B_i,b_i) are thus finitely presentable and the morphisms $\gamma_i : (B_i,b_i) \to (A,a)$ constitute an extremal epimorphic family of morphisms with target (A,a). It follows that the set of finitely presentable objects of $\mathbb{A}^{\mathbb{L}}$ properly generates the category $\mathbb{A}^{\mathbb{L}}$, and thus that $\mathbb{A}^{\mathbb{L}}$ is a locally finitely presentable category. ■

On says that a comonad, $\mathbb{L} = (L,\varepsilon,\delta)$ preserves finite products if the functor, L, preserves them.

1.12.2. Proposition. *If $\mathbb{A}$ is a locally indecomposable category and $\mathbb{L}$ is a comonad of finite rank on $\mathbb{A}$, preserving finite products, the category $\mathbb{A}^{\mathbb{L}}$ of $\mathbb{L}$-coalgebras is locally indecomposable.*

Proof : The category $\mathbb{A}^{\mathbb{L}}$ is locally finitely presentable (Proposition 1.12.1) and the forgetful functor $U^{\mathbb{L}} : \mathbb{A}^{\mathbb{L}} \to \mathbb{A}$ creates finite products. On the initial object, Z, of $\mathbb{A}$, there is a unique $\mathbb{L}$-coalgebra structure which makes Z, the initial object of $\mathbb{A}^{\mathbb{L}}$. The object $Z^2 = Z \times Z$ of $\mathbb{A}^{\mathbb{L}}$ is finitely presentable in $\mathbb{A}^{\mathbb{L}}$, since its image under $U^{\mathbb{L}}$ is finitely presentable in $\mathbb{A}$. As the functor $U^{\mathbb{L}}$ preserves and reflects finite products and amalgamated sums, finite products are codisjoint and couniversal in $\mathbb{A}^{\mathbb{L}}$ as they were in $\mathbb{A}$. The category, $\mathbb{A}^{\mathbb{L}}$ is thus locally indecomposable. ∎

1.12.3. Example. *The category $\mathbb{A}\text{lgc}\mathbb{I}\text{nt}(R)$ of commutative R-algebras integral over a commutative ring R is comonadic for an idempotent comonad of finite rank which preserves finite products, on the category $\mathbb{A}\text{lgc}(R)$ of commutative R-algebras.*

The category $\mathbb{A}\text{lgc}\mathbb{I}\text{nt}(R)$ is the full subcategory of the category $\mathbb{A}\text{lgc}(R)$, whose objects are the R-algebras, which are integral over R, i.e. the R-algebras whose elements are integral over R. For each R-algebra A, denote by LA, the integral closure of R in A, i.e. the set of elements of A integral over R. LA is a sub-R-algebra of A and each R-algebra morphism $f : A \to B$ induces a morphism of R-algebras, $Lf : LA \to LB$. This defines an idempotent endofunctor $L : \mathbb{A}\text{lgc}(R) \to \mathbb{A}\text{lgc}(R)$ and the inclusion of LA into A defines a natural transformation $\varepsilon : L \to 1_{\mathbb{A}\text{lgc}(R)}$, which gives L the structure of an idempotent comonad, $\mathbb{L}$, on $\mathbb{A}\text{lgc}(R)$. The category $\mathbb{A}\text{lgc}(R)^{\mathbb{L}}$ is equivalent to the category $\mathbb{A}\text{lgc}\mathbb{I}\text{nt}(R)$. Let $(\iota_i : A_i \to A)_{i \in \mathbb{I}}$ be a filtered colimit of R-algebras. For each element α of LA, there is an object i of $\mathbb{I}$, an element α_i of A_i and a unitary polynomial $P(X) \in R[X]$, satisfying $\iota_i(\alpha_i) = \alpha$ and $P(\alpha) = 0$. The relation $\iota_i(P(\alpha_i)) = P(\iota_i(\alpha_i)) = P(\alpha) = 0$ implies the existence of a morphism $u : i \to j$ in $\mathbb{I}$ such that $A_u(P(\alpha_i)) = 0$. Then the element $\alpha_j = A_u(\alpha_i)$ of A_j belongs to LA ; since $P(\alpha_j) = P(A_u(\alpha_i)) = A_u(P(\alpha_i)) = 0$ and its image under ι_j is $\iota_j(\alpha_j) = \iota_j(A_u(\alpha_i)) = \iota_i(\alpha_i) = \alpha$. This proves that the functor L preserves filtered colimits. Let $f : A \to B$ be a morphism from a finitely presentable R-algebra, A, to an integral R-algebra, B. The R-algebra A is of the form $R[\alpha_1,\ldots,\alpha_n]$. As the elements $f(\alpha_1),\ldots,f(\alpha_n)$, are integral over R, there exist unitary polynomials $P_1(X),\ldots,P_n(X)$ of $R[X]$ satisfying $P_1(f(\alpha_1)) = \ldots = P_n(f(\alpha_n)) = 0$.

Let C be the quotient R-algebra of A by the ideal generated by the elements $P_1(\alpha_1),\ldots,P_n(\alpha_n)$. It is finitely presentable and integral as it is generated by the integral elements $\bar{\alpha}_1,\ldots,\bar{\alpha}_n$, the cosets corresponding to $\alpha_1,\ldots,\alpha_n$. The morphism f factorises as the composite of the quotient $q : A \to C$, and an R-algebra morphism $C \to B$. It thus follows that the comonad $\mathbb{L}$ is of finite rank. Since it preserves finite products and as the category $\mathbb{A}lgc(R)$ is locally indecomposable, the category $\mathbb{A}lgc\mathbb{I}nt(R)$ is locally indecomposable. ∎

CHAPTER II

CATEGORIES OF BOOLEAN SHEAVES OF SIMPLE ALGEBRAS

2.1. LOCALLY SIMPLE CATEGORIES.

An object A of a category $\mathbb{A}$, detects monomorphisms if the functor $\mathrm{Hom}_{\mathbb{A}}(A,-)$ reflects monomorphisms, that is, if any morphism f of $\mathbb{A}$ such that the map $\mathrm{Hom}_{\mathbb{A}}(A,f)$ is injective, is necessarily a monomorphism.

2.1.0. Definition. A category is locally simple if

(1) it is locally finitely presentable,

(2) the product of its initial object with itself is finitely presentable and detects monomorphisms,

(3) its finite products are codisjoint and couniversal.

A locally simple category is locally indecomposable (1.1.1). We will show that locally simple categories are characterised amongst the locally indecomposable categories by the property of having enough direct factor quotient objects.

2.1.1. Notations. The object A is said to be simple if it has exactly two regular quotient objects i.e., if $R(A) \simeq \{0,1\} = 2$. The object, A, is said to be projective if, for any regular epimorphism f of $\mathbb{A}$, the map $\mathrm{Hom}_{\mathbb{A}}(A,f)$ is surjective. It is said to be a generator for regular epimorphisms [21, 1.7.1] if any regular epimorphism, f, such that the map $\mathrm{Hom}_{\mathbb{A}}(A,f)$ is bijective, is necessarily an isomorphism. The object A is decidable [23, p. 162] if the diagonal subobject $\Delta_A : A \to A\times A$ is complemented. Dually, A is codecidable if the codiagonal quotient object $\Delta_A : A \coprod A \to A$ is complemented in $Q(A \coprod A)$.

2.1.2. Proposition. For a locally indecomposable category, $\mathbb{A}$, the following assertions are equivalent :

(0) $\mathbb{A}$ is locally simple.

(1) Any regular quotient object is a cointersection of

direct factors.

(2) *Any finitely generated regular quotient object is a direct factor*.

(3) *Any finitely generated regular quotient object is complemented*.

(4) *Any finitely presentable object is codecidable*.

(5) *The object Z^2 is a projective generator for regular epimorphisms*.

(6) *Any indecomposable object is simple*.

Proof : *(0) => (1)*. Let $q : A \to Q$ be a regular epimorphism. The set $\mathcal{V} = \{\delta \in \Delta(A) : \delta$ factorises $q\} = \Delta(q)^{-1}(\{1_Q\})$ is a filter in the Boolean algebra $\Delta(A)$. Let $m : A \to M$ be the cointersection of the quotient objects $\delta \in \mathcal{V}$. The morphism q factorises in the form $q = nm$ and $\Delta(m) \simeq \varinjlim_{\delta\in\mathcal{V}}\Delta(\delta) \simeq \varinjlim_{\delta\in\mathcal{V}}\Delta(A)/(\delta) \simeq \Delta(A)/\mathcal{V}$, is the coimage of $\Delta(q)$. Consequently $\Delta(n)$ is a monomorphism and n is a monomorphism, therefore an isomorphism. As a result, $q \simeq m$ is a cointersection of direct factors of A.

(1) => (2). Let $q : A \to Q$ be a coequaliser of a pair of morphisms $(m,n) : X \rightrightarrows A$ with X finitely presentable. The quotient object q, being a cointersection of direct factors of A, is a cointersection of elements of $\mathcal{V} = \Delta(q)^{-1}$ $(\{1_Q\})$ and consequently, $q = \varinjlim_{\delta\in\mathcal{V}}\delta$ is a filtered colimit. The equality $qm = qn$ implies the existence of a $\delta \in \mathcal{V}$ satisfying $\delta m = \delta n$, so $\delta \simeq q$.

(2) <=> (3). The direct factors are precisely the complemented quotient objects, (Proposition 1.3.2).

(3) => (4). For any finitely presentable object X, the codiagonal $\nabla_X : X \coprod X \to X$ is a finitely generated regular quotient object, and therefore is a complemented quotient object.

(4) => (3). A coequaliser of a pair of morphisms $(m,n) : X \rightrightarrows A$ with X finitely presentable is a direct image of the codiagonal $\nabla_X : X \coprod X \to X$ by the morphism $\langle m,n\rangle : X \coprod X \to A$; it is therefore a direct factor (Proposition 1.3.2).

(2) => (5). Any regular epimorphism $q : A \to Q$ is an object of the category $A/\mathbb{A}$, and is a filtered colimit of finitely generated regular epimorphisms ; say $q = \varinjlim_{i\in\mathbb{I}} q_i$. As each morphism q_i is a direct factor of A, each homomorphism $\Delta(q_i)$ is surjective and it follows that the homomorphism $\Delta(q)$ is surjective. The same is thus true of the map $\mathrm{Hom}_{\mathbb{A}}(Z^2,q)$. The object Z^2 is therefore projective.

If the map $\mathrm{Hom}_{\mathbb{A}}(Z^2,q)$ is bijective, then, for each $i \in \mathbb{I}$, the map $\mathrm{Hom}_{\mathbb{A}}(Z^2,q_i)$ is injective, therefore bijective and consequently, $\Delta(q_i)$ is bijective, so q_i is an isomorphism, which implies that q is an isomorphism. The object Z^2 is therefore a generator for regular epimorphisms.

(5) => (6). Let A be an indecomposable object. As the object A is not isomorphic to the final object 1, the two regular quotient objects 1_A and 0_A are distinct. If $q : A \to Q$ is a regular quotient object of A, the function $\mathrm{Hom}_{\mathbb{A}}(Z^2,q)$ is surjective and therefore the homomorphism $\Delta(q) : \Delta(A) \to \Delta(Q)$ is surjective. Since $\Delta(A) \simeq 2$, $\Delta(Q) \simeq 2$ or $\Delta(Q) \simeq \{0\}$. It follows that $\Delta(q)$ is either an isomorphism, in which case q is an isomorphism and $q \simeq 1_A$, or $Q \simeq 1$, so $q \simeq 0_A$. The object A is therefore simple.

(6) => (0). First, let us note that any morphism $f : X \to Y$ between two simple objects is monomorphic : if $(m,n) : M \rightrightarrows X$ is a pair of morphisms satisfying $fm = fn$, the coequaliser $q : X \to Q$ of (m,n) is such that $q \simeq 0_X$ or $q \simeq 1_X$, but $q \neq 0_X$ as $Y \neq 1$, so $q \simeq 1_X$ and $m = n$. Further, we note that, for any object A of $\mathbb{A}$, the family $(\eta_\Psi : A \to A_\psi)_{\Psi\in \mathrm{Spec}_{\mathrm{Ind}}(A)}$ of indecomposable components of A is monomorphic and consequently the morphism $(\eta_\Psi) : A \to \prod_{\Psi\in\mathrm{Spec}_{\mathrm{Ind}}(A)} A_\psi$ is a monomorphism (Proposition 1.7.2). Let us consider now a morphism $f : A \to B$ such that the map $\mathrm{Hom}_{\mathbb{A}}(Z^2,f)$ is injective. The homomorphism $\Delta(f) : \Delta(A) \to \Delta(B)$ is injective, and the function $\mathrm{Spec}_{\mathrm{Ind}}(f) : \mathrm{Spec}_{\mathrm{Ind}}(B) \to \mathrm{Spec}_{\mathrm{Ind}}(A)$ is surjective, which implies that the canonical morphism $d : \prod_{\Psi\in\mathrm{Spec}_{\mathrm{Ind}}(A)} A_\psi \longrightarrow \prod_{\Phi\in\mathrm{Spec}_{\mathrm{Ind}}(B)} A_{\mathrm{Spec}_{\mathrm{Ind}}(f)(\Phi)}$ is injective. As the indecomposable components of A and B are simple objects, any morphism between these objects is a monomorphism and, in particular, for each $\Phi \in \mathrm{Spec}_{\mathrm{Ind}}(B)$, the morphism $f_\Phi : A_{\mathrm{Spec}_{\mathrm{Ind}}(f)(\Phi)} \to B_\Phi$ satisfying $f_\Phi \eta_{\mathrm{Spec}_{\mathrm{Ind}}(f)(\Phi)} = \eta_\Phi f$ is a monomorphism. The product morphism $\prod_{\Phi\in\mathrm{Spec}_{\mathrm{Ind}}(B)} f_\Phi : \prod_{\Phi\in\mathrm{Spec}_{\mathrm{Ind}}(B)} A_{\mathrm{Spec}_{\mathrm{Ind}}(f)(\Phi)} \to \prod_{\Phi\in\mathrm{Spec}_{\mathrm{Ind}}(B)} B_\Phi$ is therefore a monomorphism. It follows that the composite morphism $(\prod_{\Phi\in\mathrm{Spec}_{\mathrm{Ind}}(B)} f_\Phi) d (\eta_\psi)_{\psi\in\mathrm{Spec}_{\mathrm{Ind}}(A)} : A \to \prod_{\Phi\in\mathrm{Spec}_{\mathrm{Ind}}(B)} B_\Phi$ is a monomorphism ; but this morphism is equal to the morphism

$(\eta_\Phi)f : A \to \prod_{\Phi \in Spec(B)} B_\Phi$. It follows that f is a monomorphism. The object Z^2 thus detects monomophisms, and the category $\mathbb{A}$ is locally simple. ∎

2.2. EXAMPLES OF LOCALLY SIMPLE CATEGORIES.

All the ring and algebras considered are unitary. We use the notations of 1.2.

2.2.0. $\mathbb{R}ng\mathbb{S}t\mathbb{R}eg$: The category of strongly regular rings. This category is locally indecomposable and any finitely generated ideal of a strongly regular ring A is generated by a central idempotent. Therefore any finitely generated regular quotient object of A is a direct factor of A in $\mathbb{R}ng\mathbb{S}t\mathbb{R}eg$.

2.2.1. $\mathbb{R}ngc\mathbb{R}eg$: The category of regular commutative rings.

2.2.2. $\mathbb{R}ngc\mathbb{R}eg(p)$: Regular commutative rings of prime characteristic p.

2.2.3. p-$\mathbb{R}ngc$: The category of p-rings.

2.2.4. p^k-$\mathbb{R}ngc$: The category of p^k-rings.

2.2.5. $\mathbb{R}ngc\mathbb{R}eg\mathbb{F}orm\mathbb{R}l$: regular commutative rings which are formally real i.e. such that $1 + X_1^2 + \dots + X_n^2$ is invertible.

2.2.6. $\mathbb{R}ngc\mathbb{R}eg\mathbb{O}rd$: Ordered regular commutative rings with positive squares.

2.2.7. $\mathbb{R}ngc\mathbb{R}eg\mathbb{S}t\mathbb{L}at$: Strongly lattice-ordered regular commutative rings i.e. regular f-rings.

2.2.8. $\mathbb{R}ngc\mathbb{B}a$: Commutative Baer rings [22].
A commutative Bäer ring is defined as being a commutative ring in which the annihilator of each element is the ideal generated by an idempotent. This amounts to associating to each element x, an idempotent e(x) satisfying $xe(x) = x$ and $(yx = 0 \Rightarrow ye(x) = 0)$. With homomorphisms of rings preserving these associated idempotents, they constitute the category $\mathbb{R}ngc\mathbb{B}a$. It is a locally finitely presentable category. The forgetful functor $U : \mathbb{R}ngc\mathbb{B}a \to \mathbb{R}ngc$ has a left adjoint and preserves filtered colimits. It lifts direct factors, since, if we have e an idempotent of a Baer ring A, the projection $A \to A/Ae$ is, in a unique way, a homomorphism of Baer rings and it is a direct factor, in $\mathbb{R}ngc\mathbb{B}a$, which is a strongly cocartesian morphism for U. The functor U is

therefore a morphism of locally indecomposable categories (Proposition 1.8.3). Moreover, any regular quotient object in $\mathbb{R}ngc\mathbb{B}a$ is of the form $A \to A/I$, where I is an ideal of A satisfying $(x \in I \Rightarrow e(x) \in I)$, and therefore generated by idempotents. Consequently, the quotient object $A \to A/I$ is a counion of direct factors. The category $\mathbb{R}ngc\mathbb{B}a$ is therefore locally simple (Proposition 2.1.2).

2.2.9. <u>$\mathbb{R}ngc\mathbb{B}a\mathbb{O}rd$</u> : Ordered, commutative Baer ring with positive squares.

2.2.10. <u>$\mathbb{R}ngc\mathbb{B}a\mathbb{L}at$</u> : Lattice-Ordered, commutative Baer rings with positive squares.

2.2.11. <u>$\mathbb{R}ngc\mathbb{B}a\mathbb{S}t\mathbb{L}at$</u> : Strongly lattice-ordered commutative Baer rings.

2.2.12. <u>$\mathbb{R}ngc\mathbb{S}t\mathbb{L}at\mathbb{P}roj$</u> : Projectable, strongly lattice-ordered commutative rings [24].
They are defined as being strongly lattice-ordered commutative rings in which any principal polar ideal is a direct factor. This amounts to associating to each element x, an idempotent $e(x)$ satisfying $xe(x) = x$ and $(|y| \wedge |x| = 0 \Rightarrow ye(x) = 0)$. With homomorphisms of lattice-ordered rings preserving associated idempotents, they make up the category $\mathbb{R}ngc\mathbb{S}t\mathbb{L}at\mathbb{P}roj$. One shows that the forgetful functor $U : \mathbb{R}ngc\mathbb{S}t\mathbb{L}at\mathbb{P}roj \to \mathbb{R}ngc\mathbb{S}t\mathbb{L}at$ is a morphism of locally indecomposable categories so as to deduce that $\mathbb{R}ngc\mathbb{S}t\mathbb{L}at\mathbb{P}roj$ is locally simple.

2.2.13. <u>$\mathbb{R}ng\ \mathbb{E}$</u> : Category of E-rings [26].
An E-ring is defined by Kennison as being a ring A with a unary operation $e : A \to A$ satisfying the six axioms :

(E1) $e(x)$ is a central idempotent
(E2) $x\,e(x) = x$
(E3) $e(e(x)) = e(x)$
(E4) $e(-x) = e(x)$
(E5) $e(1-e(x)) = 1 - e(x)$
(E6) $e(e(x)y) = e(x)\,e(y)$

One can see that half of these axioms are not needed and that the following four are sufficient :

(1) $e(x)$ is a central element
(2) $e(0) = 0$
(3) $x\,e(x) = x$
(4) $e(y\,e(x)) = e(y)\,e(x)$.

With homomorphisms of rings which preserve the operation e, they constitute the category $\mathbb{E}$-$\mathbb{R}$ng.

2.2.14. $\mathbb{R}$ngc $\mathbb{E}$: Category of commutative E-rings.

2.2.15. $\mathbb{R}$ngc$\mathbb{E}$$\mathbb{O}$rd : Ordered commutative E-rings with positive squares.

2.2.16. $\mathbb{R}$ngc$\mathbb{E}$$\mathbb{S}t\mathbb{L}$at : Commutative, strongly lattice-ordered E-rings.

2.2.17. $\mathbb{R}$ngc$\mathbb{D}$if$\mathbb{R}$eg : Regular differential rings.

2.2.18. $\mathbb{R}$ngc$\mathbb{D}$if$\mathbb{R}$eg(p) : Regular differential rings of prime characteristic p.

2.2.19. $\mathbb{R}$ngc$\mathbb{D}$if$\mathbb{R}$eg$\mathbb{P}$erf(p) : Regular differential rings which are differentially perfect of prime characteristic, p, [28].

2.2.20. $\mathbb{B}$ool : The category of Boolean algebras.

2.2.21. Any variety of any of the above categories i.e. any full subcategory having as objects those that satisfy a set of identities.

2.2.22. Any category of algebras whose underlying rings constitute one of the above listed categories.

2.2.23. Any finite product of the above mentioned categories.

2.3. LOCALLY SIMPLE CATEGORIES OF FUNCTORS.

2.3.0. Theorem. If $\mathbb{C}$ is a small finitely complete category with disjoint and universal finite sums, and with decidable objects, the category, $\mathbb{C}\mathrm{ont}_{\aleph_0}[\mathbb{C},\mathbb{E}\mathrm{ns}]$, of finitely continuous functors defined on $\mathbb{C}$ with values in $\mathbb{E}$ns, is locally simple. Its simple objects are the finitely continuous functors which preserve finite sums.

Proof : We denote $\mathbb{A}$, the category $\mathbb{C}\mathrm{ont}_{\aleph_0}[\mathbb{C},\mathbb{E}\mathrm{ns}]$. It is a locally finitely presentable category and the natural Yoneda embedding, $\mathbb{C}^{op} \to \mathbb{A}$ preserves finite colimits and finite products. This embedding induces an equivalence between the category $\mathbb{C}^{op}$ and the full subcategory of $\mathbb{A}$, whose objects are the finitely presentable objects [16]. The initial object of $\mathbb{A}$ is the functor $\mathrm{Hom}_{\mathbb{C}}(1,-)$ and its square $(\mathrm{Hom}_{\mathbb{C}}(1,-))^2 \simeq \mathrm{Hom}_{\mathbb{C}}(1 \coprod 1,-)$ is a finitely presentable object. If one denote by 0, the strict initial object of $\mathbb{C}$, the functor $\mathrm{Hom}_{\mathbb{C}}(0,-)$

is the final object of $\mathbb{A}$. It is the constant functor of value 1. Let us show that it is a strict final object. Let $\alpha : \mathrm{Hom}_{\mathbb{C}}(0,-) \to F$ be a morphism of $\mathbb{A}$. The existence of the function $\alpha_0 : \mathrm{Hom}_{\mathbb{C}}(0,0) \to F(0)$ implies that the set $F0$ is not empty. The functor F is a filtered colimit of representable functors : say $F = \varinjlim_{i\in\mathbb{I}} \mathrm{Hom}_{\mathbb{C}}(X_i,-)$. Then the set $\varinjlim_{i\in\mathbb{I}} \mathrm{Hom}_{\mathbb{C}}(X_i,0)$ is non-empty, which implies that there exists an object i_0 of $\mathbb{I}$ such that, for all the objects i of $\mathbb{I}$ below i_0, the set $\mathrm{Hom}_{\mathbb{C}}(X_i,0)$ is non-empty, that is $X_i \simeq 0$. It follows that $F = \varinjlim_{i\in\mathbb{I}} \mathrm{Hom}_{\mathbb{C}}(0,-) \simeq \mathrm{Hom}_{\mathbb{C}}(0,-)$.

Let F,G,H be three objects of $\mathbb{A}$. There exists a filtered category $\mathbb{I}$ and three diagrams $(X_i)_{i\in\mathbb{I}}$, $(Y_i)_{i\in\mathbb{I}}$, $(T_i)_{i\in\mathbb{I}}$ of $\mathbb{C}$ indexed by $\mathbb{I}$ such that $F = \varinjlim_{i\in\mathbb{I}} \mathrm{Hom}_{\mathbb{C}}(X_i,-)$, $G = \varinjlim_{i\in\mathbb{I}} \mathrm{Hom}_{\mathbb{C}}(Y_i,-)$, $H = \varinjlim_{i\in\mathbb{I}} \mathrm{Hom}_{\mathbb{C}}(T_i,-)$. For each $i \in \mathbb{I}$, denote by $(r_i : X_i \to X_i \coprod Y_i$, $s_i : Y_i \to X_i \coprod Y_i)$, the sum of X_i and Y_i. Then the pair $(\mathrm{Hom}_{\mathbb{C}}(r_i,-) : \mathrm{Hom}_{\mathbb{C}}(X_i \coprod Y_i,-) \to \mathrm{Hom}_{\mathbb{C}}(X_i,-)$, $\mathrm{Hom}_{\mathbb{C}}(s_i,-) : \mathrm{Hom}_{\mathbb{C}}(X_i \coprod Y_i,-) \to \mathrm{Hom}_{\mathbb{C}}(Y_i,-))$ is a product. The same is true of the pair $(\varinjlim_{i\in\mathbb{I}} \mathrm{Hom}_{\mathbb{C}}(r_i,-) : \varinjlim_{i\in\mathbb{I}} \mathrm{Hom}_{\mathbb{C}}(X_i \coprod Y_i,-) \longrightarrow \varinjlim_{i\in\mathbb{I}} \mathrm{Hom}_{\mathbb{C}}(X_i,-)$, $\varinjlim_{i\in\mathbb{I}} \mathrm{Hom}_{\mathbb{C}}(s_i,-) : \varinjlim_{i\in\mathbb{I}} \mathrm{Hom}_{\mathbb{C}}(X_i \coprod Y_i,-) \longrightarrow \varinjlim_{i\in\mathbb{I}} \mathrm{Hom}_{\mathbb{C}}(Y_i,-))$, which is the product $(p : F\times G \to F,\ q : F\times G \to G)$. For each $i \in \mathbb{I}$, the fibred product of (r_i,s_i) is the object 0, therefore the amalgamated sum of $(\mathrm{Hom}_{\mathbb{C}}(r_i,-), \mathrm{Hom}_{\mathbb{C}}(s_i,-))$ is $\mathrm{Hom}_{\mathbb{C}}(0,-)$ and, by passing to the filtered colimit over $\mathbb{I}$, the amalgamated sum of (p,q) is $\mathrm{Hom}_{\mathbb{C}}(0,-)$. Consequently, finite products are codisjoint in $\mathbb{A}$. Let us denote by $\mathbb{D}$, the category of diagrams of $\mathbb{A}$ of the form $F\times G \to H$, in which F,G,H are in $\mathbb{A}$. $\mathbb{D}$ is a locally finitely presentable category [16]. The functor $L : \mathbb{D} \to \mathbb{A}$ defined by $L(F\times G \to H) = F$ has a right adjoint $U : \mathbb{A} \to \mathbb{D}$ defined by $U(F) = F\times 1 \to 1$, which preserves the filtered colimits ; the functor L preserves, therefore, the finitely presentable objects. Similarly, the functor, $M : \mathbb{D} \to \mathbb{A}$, defined by $M(F\times G \to H) = G$, preserves the finitely presentable objects. In the same way, the functor $N : \mathbb{D} \to \mathbb{A}$ defined by $N(F\times G \to H) = H$ has a right adjoint $V : \mathbb{A} \to \mathbb{D}$ defined by $V(H) = (H\times 1 \to H)$, which preserves filtered colimits and, consequently, N preserves the finitely presentable objects. It follows that the finitely presentable objects of $\mathbb{D}$ are isomorphic to the objects of the form $\mathrm{Hom}_{\mathbb{C}}(X,-)\times\mathrm{Hom}_{\mathbb{C}}(Y,-) \to \mathrm{Hom}_{\mathbb{C}}(T,-)$ with $X,Y,T, \in \mathbb{C}$, that is, to the objects of the form $\mathrm{Hom}_{\mathbb{C}}(f,-)$ with $f : T \to X \coprod Y$ in $\mathbb{C}$. Any object

$\alpha : F\times G \to H$ of $\mathbb{D}$ is therefore of the form $\alpha = \varinjlim_{i\in\mathbb{I}} \mathrm{Hom}_{\mathbb{C}}(f_i,-)$ with $f_i : T_i \to X_i \coprod Y_i$. If m_i is the inverse image of r_i by f_i and n_i is the inverse image of s_i by f_i, the pair (m_i,n_i) is a sum. The pair $(\mathrm{Hom}_{\mathbb{C}}(m_i,-), \mathrm{Hom}_{\mathbb{C}}(n_i,-))$ is therefore a product in $\mathbb{A}$, as is the pair $(\varinjlim_{i\in\mathbb{I}} \mathrm{Hom}_{\mathbb{C}}(m_i,-), \varinjlim_{i\in\mathbb{I}} \mathrm{Hom}_{\mathbb{C}}(n_i,-))$. The direct image of p by α is $\varinjlim_{i\in\mathbb{I}} \mathrm{Hom}_{\mathbb{C}}(m_i,-)$ and the direct image of q by α is $\varinjlim_{i\in\mathbb{I}} \mathrm{Hom}_{\mathbb{C}}(n_i,-)$. The direct images therefore constitute a product, which proves that finite products are couniversal in $\mathbb{A}$.

The category $\mathbb{A} = \mathbb{C}\mathrm{ont}_{\aleph_0}[\mathbb{C},\mathbb{E}\mathrm{ns}]$ is therefore locally indecomposable. Let X be an object of $\mathbb{C}$ and $\Delta' : X' \to X\times X$, a subobject of $X\times X$, complementary to the diagonal subobject, $\Delta : X \to X\times X$. Let us show that $(\Delta : X \to X\times X,\ \Delta' : X' \to X\times X)$ is a sum in $\mathbb{C}$. Let $(i : X \to X \coprod X',\ i' : X' \to X \coprod X')$ be the sum of X and X', and $\langle\Delta,\Delta'\rangle : X \coprod X' \to X\times X$, the morphism defined by $\langle\Delta,\Delta'\rangle i = \Delta$ and $\langle\Delta,\Delta'\rangle i' = \Delta'$. The fibred product of the pair (Δ,Δ) is $(1_X,1_X)$. Denoting by 0, the initial object of $\mathbb{C}$, the fibred product of the pair (Δ,Δ') is $(0 \to X,\ 0 \to X')$. It follows from the universality of finite sums that the fibred product of the pair $(\Delta,\langle\Delta,\Delta'\rangle)$ is $(1_X,i)$. Similarly, the fibred product of $(\Delta',\langle\Delta,\Delta'\rangle)$ is $(1_X,i')$. It also follows from the universality of finite sums that the fibred product of $\langle\Delta,\Delta'\rangle$, with itself, is $(1_{X \coprod X},1_{X \coprod X'})$, that is, that the morphism $\langle\Delta,\Delta'\rangle$ is a monomorphism. The subobject $\langle\Delta,\Delta'\rangle$ of $X\times X$ is necessarily, $1_{X\times X}$, therefore, $\langle\Delta,\Delta'\rangle$ is an isomorphism and the pair $(\Delta : X \to X\times X,\ \Delta' : X' \to X\times X)$ is a sum in $\mathbb{C}$. The pair $(\mathrm{Hom}_{\mathbb{C}}(\Delta,-) : \mathrm{Hom}_{\mathbb{C}}(X\times X,-) \to \mathrm{Hom}_{\mathbb{C}}(X,-),\ \mathrm{Hom}_{\mathbb{C}}(\Delta',-) : \mathrm{Hom}_{\mathbb{C}}(X\times X,-) \to \mathrm{Hom}_{\mathbb{C}}(X',-))$ is therefore a product in $\mathbb{A}$. The codiagonal quotient object of the object $\mathrm{Hom}_{\mathbb{C}}(X,-)$ of $\mathbb{A}$, being isomorphic to the quotient object $\mathrm{Hom}_{\mathbb{C}}(\Delta,-) : \mathrm{Hom}_{\mathbb{C}}(X\times X,-) \to \mathrm{Hom}_{\mathbb{C}}(X,-)$ is therefore complemented. As any finitely presentable object of $\mathbb{A}$ is of the form, $\mathrm{Hom}_{\mathbb{C}}(X,-)$, it follows that the finitely presentable objects of $\mathbb{A}$ are codecidable and that the category $\mathbb{A}$ is locally simple (Proposition 2.1.2). Moreover, the simple objects of $\mathbb{A}$ coincide with the indecomposable objects of $\mathbb{A}$ (Proposition 2.1.2). A functor $F : \mathbb{C} \to \mathbb{E}\mathrm{ns}$, which is an indecomposable object of $\mathbb{A}$, preserves finite sums, since it is isomorphic to the composite of the canonical functor $\mathbb{C} \to \mathbb{A}^{op}$ and the functor $\mathrm{Hom}_{\mathbb{A}}(-,F) : \mathbb{A}^{op} \to \mathbb{E}\mathrm{ns}$ and they both preserve finite sums (Proposition 1.5.1). Conversely, a functor $F : \mathbb{C} \to \mathbb{E}\mathrm{ns}$, an object of $\mathbb{A}$, which preserves finite sums, is indecomposable since

$\mathrm{Hom}_{\mathbb{A}}(\mathrm{Hom}_{\mathbb{C}}(1 \coprod 1,-),F) \simeq F(1 \coprod 1) \simeq F1 \coprod F1 \simeq 1 \coprod 1$ therefore $\Delta_{\mathbb{A}}(F) = 2$. ∎

2.3.1. Examples. If $\mathbb{C}$ is a small Boolean topos [23] such as, for instance, the topos of finite cardinals of a topos [23.6.2], or a small atomic topos [3], then the category $\mathrm{Cont}_{\aleph_0}[\mathbb{C},\mathbb{E}\mathrm{ns}]$ is locally simple.

2.3.2. Theorem. Any locally simple category is equivalent to a category $\mathrm{Cont}_{\aleph_0}[\mathbb{C},\mathbb{E}\mathrm{ns}]$ of finitely continuous set valued functors defined on a small category $\mathbb{C}$ which is finitely complete with disjoint universal finite sums and with decidable objects.

Proof : Let $\mathbb{A}$ a locally simple category. The full subcategory $\mathbb{A}_{\aleph_0}$ of $\mathbb{A}$, whose objects are the finitely presentable objects of $\mathbb{A}$, is finitely cocomplete, has finite products which are codisjoint and couniversal, (Proposition 1.4.0), and its objects are codecidable, (Proposition 2.1.2). The dual category $\mathbb{C} = \mathbb{A}_{\aleph_0}^{op}$ is, therefore a finitely complete small category with disjoint, universal finite sums and with decidable objects. The theorem is then derivable from the fact that the category $\mathrm{Cont}_{\aleph_0}[\mathbb{C},\mathbb{E}\mathrm{ns}]$ is equivalent to $\mathbb{A}$, [16]. ∎

2.4. THE ALGEBRAIC NATURE OF LOCALLY SIMPLE CATEGORIES.

2.4.0. Theorem. A locally simple category is algebraic in the sense of Gabriel-Ulmer [16].

Proof : Let us consider a locally simple category $\mathbb{A}$.

a) Let A and B be two objects of $\mathbb{A}$ with product $(p : A\times B \to A,\ q : A\times B \to B)$ and $f : X \to A$, a morphism of $\mathbb{A}$. If $(m : X\times B \to X,\ n : X\times B \to B)$ is the product of A and B, then the fibred product of (p,f) is $(f\times 1_B : X\times B \to A\times B,\ m : X\times B \to X)$. Direct factors are thus universal in $\mathbb{A}$. Any regular epimorphism being a filtered colimit of direct factors, it follows that regular epimorphisms are also universal in $\mathbb{A}$ and consequently that the category $\mathbb{A}$ is regular [2,1.3.]. Let $(m,n) : R \rightrightarrows A$ be an equivalence relation on A. The morphisms m and n are split epimorphisms, therefore they are regular and hence are filtered cointersections of direct factors of R :

say $m = \wedge_{i\in I} m_i$ and $n = \wedge_{i\in I} n_i$. The pair of morphisms with source R, $(m : R \to A,\ n : R \to A)$ is monomorphic as is each pair (m_i, n_i). It follows that for each $i \in I$, we have $n_i \vee m_i = 1_R$ and (n_i, m_i) is fibred product of its amalgamated sum (cf. Theorem 1.7.1.). Passing to the filtered colimit over $\mathbb{I}$, it follows that the pair (m,n) is fibred product of its amalgamated sum, which is the pair $(q : A \to B,\ q : A \to B)$, in which q is the coequaliser of (m,n). The equivalence relation (m,n) is therefore effective and the category is exact [2,1.3.].

b) Let us show that, for any finitely presentable object, X, of $\mathbb{A}$, the object $Z\times X$ is projective. Let us denote $r : Z\times X \to Z$, the canonical projection and $i_B : Z \to B$, the only morphism from Z to an object B. For each projection, $p : A\times B \to A$, the function $\mathrm{Hom}_{\mathbb{A}}(Z\times X, p)$ is surjective, since any morphism $f : Z\times X \to A$ is the image of the morphism $(f, i_B r) : Z\times X \to A\times B$. So for any direct factor δ, the function $\mathrm{Hom}_{\mathbb{A}}(Z\times X, \delta)$ is surjective. If $q : A \to Q$ is a regular quotient object, then $q = \varinjlim_{i\in\mathbb{I}} \delta_i$ is a filtered colimit of direct factors of A. The object $Z\times X$ being finitely presentable (Proposition 1.4.0;), the mapping $\mathrm{Hom}_{\mathbb{A}}(Z\times X, q) = \mathrm{Hom}_{\mathbb{A}}(Z\times X, \varinjlim_{i\in\mathbb{I}}, \delta_i) \simeq \varinjlim_{i\in\mathbb{I}} \mathrm{Hom}_{\mathbb{A}}(Z\times X, \delta_i)$ is surjective. It follows that the object $Z\times X$ is projective.

c) Let G be a proper generating set for $\mathbb{A}$ made up of finitely presentable objects. The objects of the form $Z\times X$, where X is any object of G, are projective and finitely presentable. Let A be an object of $\mathbb{A}$. It is a regular quotient of the sum of a family of objects of G, say $q : \coprod_{i\in I} X_i \to A$, but each object X_i is a regular quotient object of the object $Z\times X_i$ by the projection $p_i : Z\times X_i \to X_i$. Therefore $\coprod_{i\in I} p_i : \coprod_{i\in I} Z\times X_i \to \coprod_{i\in I} X_i$ is a regular quotient object and A is regular quotient of $\coprod_{i\in I} Z\times X_i$. As a result, the set of objects of the form $Z\times X$ with X an object of G is a proper generator of $\mathbb{A}$. The The category $\mathbb{A}$ is therefore algebraic in the sense of Gabriel-Ulmer [16]. ■

2.4.1. Corollary. A locally simple category is equivalent to an I-algebraic category in the sense of Bénabou [6] i.e. a category of families of sets with an algebraic structure defined by internal or external operations, and identities between composite operations.

Proof : See [16]. ■

2.4.2. Proposition. A locally simple category is varietal in the sense of Linton [31].

Proof : Let $\mathcal{G}$ be a proper generating set of $\mathbb{A}$ made up of projective finitely presentable objects (Theorem 2.4.0.). Let X be an object of $\mathcal{G}$. For each object Y of $\mathcal{G}$, there exists at least one morphism $f_Y : Y \to X$ since the morphism $q : X \to 1$ is a regular epimorphism, thus, the function $Hom_{\mathbb{A}}(Y,q)$ is surjective, and the set $Hom_{\mathbb{A}}(Y,X)$ is non-empty. For $Y = X$, let us take $f_X = 1_X$. Let $G = \coprod_{Y\in\mathcal{G}} Y$. It is a projective object of which X is a regular quotient, since the morphism $\langle f_Y\rangle : \coprod_{Y\in\mathcal{G}} Y \to X$ is an epimorphism split by the canonical injection $i_X : X \to G$. Any object of $\mathbb{A}$, being a regular quotient of the sum of a family of objects of $\mathcal{G}$, is then a regular quotient of the sum of a family of objects equal to G. The object G is therefore a proper projective generator of $\mathbb{A}$. As $\mathbb{A}$ is, moreover, an exact category [2], $\mathbb{A}$ is varietal in the sense of Linton [31, Proposition 3]. ■

2.4.3. Corollary. A locally simple category is monadic over the category of sets.

Proof : See [35, 21.7.]. ■

2.4.4. Proposition. A locally simple category is algebraic in the sense of Lawvere [30] if, and only if, it has a finite set of finitely presentable objects which is a proper generating set.

Proof : The condition is necessary since an algebraic category in the sense of Lawvere has a finitely presentable proper generator [30]. Conversely, let us consider a locally simple category with a finite set of finitely presentable objects $G_1,\ldots,G_n$ which is a proper generator. The objects $Z\times G_1,\ldots,Z\times G_n$ are finitely presentable projective and constitute a proper generator according to part c) of the proof of Theorem 2.4.0. The object $\coprod_{i=1}^{n} Z\times G_i$ is then finitely presentable projective and is a proper generator (proof of Theorem 2.4.2.). The category is thus algebraic in the sense of Lawvere [30]. ■

2.4.5. Notation. From now on, the term "algebraic" will be reserved exclusively to algebraic categories in the sense of Lawvere.

2.4.6. Examples. Considering some examples of locally simple categories, the algebraic aspect may seem surprising. Here are two examples that confirm it.

2.4.6.0. $\mathbb{R}$ngcBa. One can define a commutative Baer ring as being a commutative ring with a unary operation e satisfying the four axioms :

(1) $e(0) = 0$

(2) $e(x^2) = e(x)$

(3) $x\, e(x) = x$

(4) $e(xy) = e(x)\, e(y)$.

2.4.6.1. $\mathbb{R}$ngc$\mathbb{S}$t$\mathbb{L}$at$\mathbb{P}$roj. One can define a projectable commutative ring as being a strongly lattice-ordered commutative ring with a unary operation e satisfying the four axioms :

(1) $e(0) = 0$

(2) $e(|x|) = e(x)$

(3) $x\, e(x) = x$

(4) $e(|x| \wedge |y|) = e(x)\, e(y)$

2.5. EQUIVALENCE RELATIONS ON AN OBJECT.

In what follows, we consider a locally simple category $\mathbb{A}$.

2.5.0. Relations on an object.
Let A be an object of $\mathbb{A}$. The category $\mathbb{A}$ has few subobjects i.e. is well-powered [33, p.126]. We denote by S(A), the ordered set of subobjects of A and we identify a subobject with one of its representatives [33, p.122]. Since $\mathbb{A}$ is complete, S(A) is a complete lattice in which the meets are the intersections and the joins, the unions. A subobject $X \to A$ is finitely generated if the object X is finitely generated in $\mathbb{A}$ [16]. These subobjects are precisely the compact elements of S(A) [18]. Any object of $\mathbb{A}$ being a monomorphic filtered colimit of finitely generated objects, any subobject of A is a union of compact subobjects of A, so that the lattice, S(A) is compactly generated [18]. As a result, the set of relations on A, i.e. of subobjects of $A^2 = A \times A$, is a complete compactly generated

lattice, $S(A^2)$. We write $E(A)$ for the set of equivalence relations on A. It is a complete sub-meet-semi-lattice of $S(A^2)$, closed under filtered unions. Since the category $\mathbb{A}$ is exact [2], the lattice, $E(A)$, is isomorphic to the dual of the lattice, $R(A)$, of regular quotient objects of A (1.3.0.). The equivalence relation $[R]$ generated by a relation R on A is the intersection of the equivalence relations on A containing R. An equivalence relation is said to be finitely generated if it is generated by a finitely generated relation. It is said to be proper if it is distinct from 1_{A^2}.

2.5.1. Theorem. The set $E(A)$ of equivalence relations on A is a complete compactly generated Heyting algebra in which the compact elements are the finitely generated equivalence relations, and coincide with the complemented elements.

Proof : Let R_o be an equivalence relation, finitely generated by a finitely generated relation, S, on A. If $(R_i)_{i\in I}$ is a filtered family of equivalence relations on A such that $R_o \subset \bigvee_{i\in I} R_i$, then the relations $S \subset R_o \subset \bigvee_{i\in I} R_i = \bigcup_{i\in I} R_i$ imply the existence of an $i \in I$ such that $S \subset R_i$, hence such that $R_o \subset R_i$. Finitely generated equivalence relations are therefore compact. Let R be an equivalence relation on A. The relation R is, in $S(A^2)$, a filtered union of finitely generated relations on A, say $R = \bigcup_{i\in I} S_i$. The inclusions $S_i \subset R$ imply $[S_i] \subset R$, therefore $\bigcup_{i\in I}[S_i] \subset R$. Since we also have $R \subset \bigcup_{i\in I}[S_i]$, it follows that $R = \bigcup_{i\in I}[S_i] = \bigvee_{i\in I}[S_i]$. Consequently, any equivalence relation is a filtered join of finitely generated equivalence relations. Moreover, if the relation R is compact in $E(A)$, then $R = \bigvee_{i\in I}[S_i]$ implies the existence of $i \in I$ satisfying $R = [S_i]$ that is to say, that R is finitely generated. As a result, $E(A)$ is a compactly generated, complete lattice, in which the compact elements are the finitely generated equivalence relations. The isomorphism between $E(A)$ and $R(A)^{op}$ induces an isomorphism between the ordered set $E_o(A)$ of finitely generated equivalence relations on A and the dual of the ordered set $R_o(A)$ of finitely generated regular quotients of A. However $R_o(A)$ is isomorphic to the Boolean algebra $\Delta(A)$ of direct factors of A. Therefore, $E_o(A)$ is a Boolean algebra. The lattice $E(A)$ being isomorphic to the lattice of ideals of $E_o(A)$ [18, section 9, Theorem 13] is a complete Heyting algebra [18, section 9, corollary 16 and section 14, corollary 2]. ∎

2.5.2. <u>The functor $E : \mathbb{A} \to \mathbb{H}\mathrm{eytComp}$</u> .

Let $f : A \to B$ be a morphism of $\mathbb{A}$. We denote by $R(f) : R(A) \to R(B)$, the mapping which to $p : A \to P$ associates the direct image of p by f. It preserves cointersections and induces the morphism $\Delta(f) : \Delta(A) \to \Delta(B)$ (1.3.5.). Let $p,q \in R(A)$. One has $p = \bigcap_{i \in I} p_i$ and $q = \bigcap_{j \in J} q_j$ with $p_i, q_j \in \Delta(A)$. Therefore $R(f)(p \vee q) = R(f)((\bigcap_{i \in I} p_i) \vee (\bigcap_{j \in J} q_j))$

$$= R(f)(\bigcap_{(i,j) \in I \times J} (p_i \vee q_j)) = \bigcap_{(i,j) \in I \times J} R(f)(p_i \vee q_j) =$$

$$\bigcap_{(i,j) \in I \times J} \Delta(f)(p_i \vee q_j) = \bigcap_{(i,j) \in I \times J} (\Delta(f)(p_i) \vee \Delta(f)(q_j)) =$$

$$(\bigcap_{i \in I} R(f)(p_i)) \vee (\bigcap_{j \in J} R(f)(q_j)) = R(f)(\bigcap_{i \in I}(p_i)) \vee R(f)(\bigcap_{j \in J} q_j) =$$

$R(f)(p) \vee R(f)(q)$. As a result, $R(f)^{op} : R(A)^{op} \to R(B)^{op}$ is a morphism of complete Heyting algebras. However, this morphism is isomorphic to the morphism $E(f) : E(A) \to E(B)$, which associates to R, the equivalence on B generated by the regular image (i.e. obtained by regular factorisation [2]) of R by the morphism $f^2 : A^2 \to B^2$. In this way, we obtains the functor $E : \mathbb{A} \to \mathbb{H}\mathrm{eytComp}$ from $\mathbb{A}$ to the category of complete Heyting algebras.

2.6. <u>THE MAXIMAL SPECTRUM OF AN OBJECT AND ITS STRUCTURAL SHEAF</u>.

2.6.0. <u>The category of simple objects</u>.

The category $\mathbb{S}\mathrm{im}\,\mathbb{A}$ of simple objects of $\mathbb{A}$ is the full subcategory of $\mathbb{A}$, whose objects are the simple objects. It is closed in $\mathbb{A}$ under subobjects, filtered colimits and ultraproducts. (Proposition 1.5.2.). Its morphisms are monomorphisms in $\mathbb{A}$ since each of them $f : A \to B$ factorises in the form $f = mq$ with $m : C \to B$ monomorphic and $q : A \to C$ regular epimorphic, and as $B \neq 1$ implies $C \neq 1$, therefore, $q \neq 0_A$ and consequently $q \simeq 1_A$.

2.6.1. <u>Definition</u>. A <u>simple component</u> of an object A is a regular quotient object $q : A \to Q$ whose target Q is a simple object. Alternatively, it is a quotient object of A by a maximal proper equivalence relation.

2.6.2. <u>Proposition</u>. <u>The category, $\mathbb{S}\mathrm{im}\,\mathbb{A}$, of simple objects of $\mathbb{A}$ is a multireflexive subcategory of $\mathbb{A}$</u> (cf. Proposition 1.7.0).

Proof : The family of simple components of an object A is a universal family of morphisms from A to $\mathbb{S}im\, \mathbb{A}$, since any morphism $f : A \to S$ with S simple, factorises in the form $f = mq$ with $m : T \to S$ monomorphic, therefore T simple, and $q : A \to T$ regular epimorphic, therefore, a simple component of A, and such a factorisation is unique up to isomorphism. ■

2.6.3. The maximal spectrum.

According to the general notation concerning multiadjoints, the set of maximal proper equivalence relations on an object A is the spectrum of A relative to the subcategory $\mathbb{S}im\, \mathbb{A}$. We are going to call it the maximal spectrum of A and denote it $Spec_{Max}(A)$. For each morphism $f : A \to B$, the function $Spec_{Max}(f) : Spec_{Max}(B) \to Spec_{Max}(A)$ associates to R, its inverse image by $f^2 : A^2 \to B^2$. In that way, we obtain the functor $Spec_{Max} : \mathbb{A}^{op} \to \mathbb{E}ns$. The spectral topology on $Spec_{Max}(A)$ is defined in the following way. Let us denote by $D : E(A) \to P(Spec_{Max}(A))$, the function defined by $D(R) = \{M \in Spec_{Max}(A) : R \not\subset M\}$. It is strictly increasing and induces an isomorphism between the complete Heyting algebra, $E(A)$, and its image, written $\Omega(Spec_{Max}(A))$. This latter set is therefore a topology on $Spec_{Max}(A)$ called the spectral topology. For each morphism $f : A \to B$, the mapping $Spec_{Max}(f) : Spec_{Max}(B) \to Spec_{Max}(A)$ is continuous.

2.6.4. Proposition. The maximum spectrum of A is a Boolean topological space whose collection of open sets is a Heyting algebra isomorphic to the Heyting algebra, $E(A)$, of equivalence relations on A and the set of open-closed subsets is a Boolean algebra isomorphic to the Boolean algebra, $E_o(A)$ of finitely generated equivalence relations on A.

Proof : It is enough to consider the isomorphism $D : E(A) \to \Omega(Spec_{Max}(A))$ which induces an isomorphism, $D_o : E_o(A) \to \Omega_o(Spec_{Max}(A))$ between the Boolean algebra $E_o(A)$ and the algebra of open-closed subsets of $Spec_{Max}(A)$. ■

2.6.5. The structural sheaf.

The ordered set $\Omega_o(Spec_{Max}(A))$, of open-closed subsets of $Spec_{Max}(A)$ has the structure of a site [21], whose topology has, as covering families those families $(u_i \to u)_{i\in I}$ such that $u = \bigcup_{i\in I} u_i$. The functor $\tilde{A} : \Omega_o(Spec_{Max}(A))^{op} \to \mathbb{A}$, assigns to $D_o(R)$, the object $\tilde{A}(D_o(R)) = A/R'$, quotient of A by the complement of R in $E_o(R)$ and assigns to a morphism $D_o(R) \to D_o(S)$, the canonically defined morphism $\tilde{A}(D_o(R) \to D_o(S)) : A/S' \to A/R'$.

2.6.5.0. Theorem. The functor $\tilde{A} : \Omega_o(\mathrm{Spec}_{Max}(A))^{op} \to \mathbb{A}$ is a sheaf with values in $\mathbb{A}$.

Proof : We will use the notations and the results of (1.7.). One has at one's disposal an isomorphism $h : \Omega_o(\mathrm{Spec}_{Max}(A)) \simeq E_o(A) \simeq \Delta(A)^{op} \simeq \Delta(A) \simeq \mathcal{D}(\mathrm{Spec}_{Ind}(A))$ which assigns to $D_o(R)$, the set $D(q_{R'} : A \to A/R')$. The functor $\tilde{A} : \Omega_o(\mathrm{Spec}_{Max}(A))^{op} \to \mathbb{A}$ is the composite of the functor h^{op} and the sheaf $\tilde{A} : \mathcal{D}(\mathrm{Spec}_{Ind}(A)) \to \mathbb{A}$ and is therefore a sheaf. ■

The sheaf $\tilde{A}$ can be extended in a unique way to a sheaf with base, $\mathrm{Spec}_{Max}(A)$ with values in $\mathbb{A}$. Its stalks are the simple components of A and its object of global sections is A (Proposition 1.7.2). Once again, it is denoted by $\tilde{A}$ and it will be called the structural sheaf on $\mathrm{Spec}_{Max}(A)$.

2.7. BOOLEAN SHEAVES OF SIMPLE OBJECTS.

The category $\mathrm{\$hvBoolSim}\,\mathbb{A}$ of sheaves of simple objects of $\mathbb{A}$ over Boolean topological spaces has, as objects, the pairs (X,F) made up of a Boolean topological space X and of a sheaf $F : \Omega(X)^{op} \to \mathbb{A}$ with base X and values in $\mathbb{A}$, whose stalks are simple objects of $\mathbb{A}$, and it has, as morphisms $(X,F) \to (Y,G)$, the pairs (f,α) made up of a continuous mapping $f : Y \to X$ and of a natural transformation $\alpha : F \to Gf^*$. The global section functor $\Gamma : \mathrm{\$hvBoolSim}\,\mathbb{A} \to \mathbb{A}$ is defined by $\Gamma(X,F) = FX$ and $\Gamma(f,\alpha) = \alpha_X$. The structural sheaf functor $\Sigma : \mathbb{A} \to \mathrm{\$hvBoolSim}\,\mathbb{A}$ is defined by $\Sigma(A) = (\mathrm{Spec}_{Max}(A), \tilde{A})$ and $\Sigma(f) = (\mathrm{Spec}_{Max}(f), \tilde{f})$ where $(\tilde{f})_{D(R)}$ is the quotient morphism of f.

2.7.0. Theorem. Any locally simple category $\mathbb{A}$ is equivalent to the category $\mathrm{\$hvBoolSim}\,\mathbb{A}$ of sheaves of simple objects of $\mathbb{A}$ on Boolean topological spaces, and the functors $\Sigma : \mathbb{A} \to \mathrm{\$hvBoolSim}\,\mathbb{A}$ and $\Gamma : \mathrm{\$hvBoolSim}\,\mathbb{A} \to \mathbb{A}$ are equivalences of categories, quasi-inverse to each other.

Proof : It follows directly from Theorem 1.7.3. ■

2.8. MORPHISMS OF LOCALLY SIMPLE CATEGORIES.

2.8.0. Definition. A morphism of locally simple categories is a functor, $U : \mathbb{A} \to \mathbb{B}$,

(i) whose source $\mathbb{A}$ and target $\mathbb{B}$ are two locally simple categories,

(ii) which preserves filtered colimits, and

(iii) has a left adjoint preserving finite products.

2.8.1. Proposition. A morphism of locally simple categories is an exact morphism of locally indecomposable categories.

Proof : Let $U : \mathbb{A} \to \mathbb{B}$ be a morphism of locally simple categories. It is obviously a morphism of locally indecomposable categories (Definition 1.8.0.). It is a continuous functor, therefore it preserves kernel pairs and direct factors. Since any regular epimorphism is a filtered colimit of direct factors, it also preserves regular epimorphisms. It thus preserves exact sequences and so it is an exact functor [2]. ∎

2.8.2. Proposition. A functor $U : \mathbb{A} \to \mathbb{B}$ is a morphism of locally simple categories if, and only if, the category $\mathbb{B}$ is locally simple and U is a morphism of locally indecomposable categories which reflects monomorphisms.

Proof : Let $U : \mathbb{A} \to \mathbb{B}$ be a morphism of locally indecomposable categories. For each morphism f of $\mathbb{A}$, we have $\mathrm{Hom}_{\mathbb{A}}(Z^2,f) \simeq \mathrm{Hom}_{\mathbb{B}}(Z^2,Uf)$ (Proposition 1.8.1.). If U is a morphism of locally simple categories, we have : Uf monomorphic => $\mathrm{Hom}_{\mathbb{B}}(Z^2,Uf)$ injective => $\mathrm{Hom}_{\mathbb{A}}(Z^2,f)$ injective => f monomorphic ; therefore U reflects monomorphisms. Conversely, if $\mathbb{B}$ is locally simple and U reflects monomorphisms, then we have : $\mathrm{Hom}_{\mathbb{A}}(Z^2,f)$ injective => $\mathrm{Hom}_{\mathbb{B}}(Z^2,Uf)$ injective => Uf monomorphic => f monomorphic. Consequently, Z^2 dectects monomorphisms in $\mathbb{A}$, and $\mathbb{A}$ is locally simple. ∎

2.8.3. Lifting of regular quotient objects.
The notion of lifting regular quotient objects is obtained from the idea of lifting coequalisers [35, 21.3.7.], by taking into account the fact that quotient objects are equivalence classes of epimorphisms. A functor $U : \mathbb{A} \to \mathbb{B}$ lifts regular quotient objects if, for any object A of $\mathbb{A}$ and any regular epimorphism $q : UA \to Q$, there exists a regular

epimorphism $p : A \to P$ and an isomorphism $r : UP \to Q$ satisfying $q = r(Up)$. It lifts them uniquely if, in addition, for any other regular epimorphism $m : A \to M$ satisfying the above mentioned condition, there is an isomorphism $n : P \to M$ satisfying $np = m$.

2.8.4. Proposition. If $\mathbb{A}$ and $\mathbb{B}$ are two locally simple categories and $U : \mathbb{A} \to \mathbb{B}$ is a functor which preserves filtered colimits and has a left adjoint, the following assertions are equivalent :

(1) U is a morphism of locally simple categories.

(2) U lifts uniquely regular quotient objects.

(3) U lifts regular quotient objects and reflects the final object.

(4) U preserves simple objects.

Proof : (1) => (2). Let A be an object of $\mathbb{A}$ and $q : UA \to Q$ a regular epimorphism. The epimorphism q is a filtered colimit of direct factors, $q_i : UA \to Q_i$ $(i \in I)$. Each of the latter can be uniquely lifted to a direct factor $p_i : A \to B_i$ of A (Proposition 1.8.1.). The filtered colimit $p : A \to B$ of the morphisms p_i $(i \in I)$ is a regular epimorphism whose image by U is isomorphic to q. Let $m : A \to M$ be another regular epimorphism whose image under U is isomorphic to q. Any direct factor of A which factorises m, factorises p, since its image by U factorises q. Then, m factorises p through a morphism $n : M \to B$ whose image by U is an isomorphism. Since U reflects monomorphisms, n is monomorphic, therefore an isomorphism. The functor U thus uniquely lifts regular quotient objects.

(2) => (3). If A is an object of $\mathbb{A}$ such as $UA = 1$, the regular quotient object, $q : A \to 1$, is such that $Uq = 1_{UA} = U(1_A)$, therefore $q \simeq 1_A$ and $A \simeq 1$.

(3) => (1). According to Proposition 1.8.1., it is sufficient to show that U lifts direct factors. Let A be an object of $\mathbb{A}$ and $q : UA \to Q$ a direct factor of UA. There exists a regular quotient object $p : A \to P$ such that $Up = q$. It is a filtered colimit of direct factors, $p_i : A \to P_i$ $(i \in I)$, say. Then $q = Up$ is a filtered colimit of the direct factors Up_i $(i \in I)$. It follows that q is one of the direct factors, Up_i.

(1) => (4). U preserves simple objects since these objects are precisely the indecomposable objects and the latter are preserved by morphisms of locally indecomposable categories.

(4) => (1). Let $F : \mathbb{B} \to \mathbb{A}$ be the left adjoint to U and $e : FZ^2 \to Z^2$, the canonical morphism (cf. (1.8.)). For any simple object S, of $\mathbb{A}$, any morphism $Z^2 \to US$ can be factorised in a unique way through $\pi : Z^2 \to Z$ or $\pi' : Z^2 \to Z$, therefore, any morphism $FZ^2 \to S$ can be factorised in a unique way through $F\pi$ or $F\pi'$. It follows that the map $Hom_{\mathbb{A}}(e,S)$ is bijective and consequently e is an isomorphism. U is therefore a morphism of locally indecomposable categories, (Proposition 1.8.1.), therefore a morphism of locally simple categories (definition 2.8.0). ■

2.8.5. Examples.

2.8.5.0. $\mathbb{R}ngc\mathbb{R}eg\mathbb{O}rd \to \mathbb{R}ngc\mathbb{R}eg$. The forgetful functor U is a morphism of locally finitely presentable categories, i.e. it preserves filtered colimits and has a left adjoint. Let A be an object of $\mathbb{R}ngcRegOrd$ and e an idempotent of A. One defines an injective homomorphism of non unitary rings $f : A/Ae \to A$ by $f(\bar{x}) = x(1-e)$. This induces on the quotient ring, A/Ae, a structure of an ordered ring with positive squares. The canonical projection $A \to A/Ae$ is then increasing since the idempotent $1-e$ is positive or null and consequently : $x \leqslant y \Rightarrow x(1-e) \leqslant y(1-e) \Rightarrow \bar{x} \leqslant \bar{y}$. It is therefore a morphism of $\mathbb{R}ncg\mathbb{R}eg\mathbb{O}rd$. Similarly, we obtain a morphism $A \to A/A(1-e)$. The pair $(A \to A/Ae,\ A \to A/A(1-e))$ is then a product in $\mathbb{R}ngc\mathbb{R}eg\mathbb{O}rd$ since, on denoting by $\bar{\bar{x}}$, the class of x modulo $1-e$, we have : $(\bar{x} \leqslant \bar{y}$ and $\bar{\bar{x}} \leqslant \bar{\bar{y}}) \Rightarrow x = xe + x(1-e) \leqslant ye + y(1-e) = y$. As a result, the morphism $A \to A/Ae$ is a direct factor in $\mathbb{R}ngcRegOrd$. Moreover, it is immediate that these are strongly cocartesian for U. The functor U is therefore a morphism of locally indecomposable categories (Proposition 1.8.3). It is faithful, therefore it reflects monomorphisms and consequently, it is a morphism of locally simple categories (Proposition 2.8.2.).

2.8.5.1. $\mathbb{R}ngc\mathbb{B}a\mathbb{O}rd \to \mathbb{R}ngc\mathbb{B}a$. One can, show, as above, that the forgetful functor is a morphism of locally simple categories.

2.9. MONADIC MORPHISMS.

2.9.0. Creation of regular quotient objects.
The notion of creation of regular quotient objects can be obtained from that of creation of coequalisers [33, p.147] by taking account of the

fact that a quotient object is an equivalence class of epimorphisms. A functor $U : \mathbb{A} \to \mathbb{B}$ <u>creates regular quotient objects</u> if, for each object A of $\mathbb{A}$ and regular epimorphism $q : UA \to Q$,

(1) there is a morphism $p : A \to P$ and an isomorphism $r : UP \to Q$ satisfying $q = r(Up)$,

(2) for any other morphism $p_1 : A \to P_1$ satisfying condition (1), there is an isomorphism $s : P \to P_1$ satisfying $sp = p_1$,

(3) the morphism p satisfying condition (1) is a regular epimorphism.

2.9.0.0. <u>Proposition</u>. <u>A functor U creates regular quotient objects if, and only if, it reflects them and lifts them uniquely</u>.

<u>Proof</u> : Immediate from (2.8.3.). ■

2.9.0.1. <u>Proposition</u>. <u>If $\mathbb{A}$ has equalisers (resp. coequalisers) and $U : \mathbb{A} \to \mathbb{B}$ is a functor which preserves them, then U creates regular quotient objects if, and only if, it lifts them and reflects isomorphisms</u>.

<u>Proof</u> : Let $U : \mathbb{A} \to \mathbb{B}$ be a functor which creates regular quotient objects. It lifts regular quotient objects. It reflects isomorphisms since a morphism $p : A \to P$ of $\mathbb{A}$ whose image by U is an isomorphism, satisfies $(Up)^{-1}(Up) = 1_{UA} = U(1_A)$, and it follows that there is an isomorphism $r : P \to A$ satisfying $rp = 1_A$, thus p is an isomorphism. Conversely, let $U : \mathbb{A} \to \mathbb{B}$ be a functor which preserves equalisers (resp. coequalisers), lifts regular quotient objects and reflects isomorphisms. It is faithful since, if $f,g : X \rightrightarrows Y$ are two morphisms of $\mathbb{A}$ such that $Uf = Ug$, their equaliser (resp. coequaliser) has a image by U, the equaliser (resp. coequaliser) of (Uf,Ug), which is an isomorphism and thus it is itself an isomorphism, and it follows that $f = g$. Let A be an object of $\mathbb{A}$ and $q : UA \to Q$ a regular quotient object of UA. There is a regular epimorphism $p : A \to P$ and an isomorphism $r : UP \to Q$ satisfying $r(Up) = q$. Let $p_1 : A \to P_1$ be a morphism of $\mathbb{A}$ and $r_1 : UP_1 \to Q$ an isomorphism satisfying $r_1(Up_1) = q$. If $(m,n) : X \rightrightarrows A$ is a pair of morphisms of $\mathbb{A}$ whose coequaliser is p, the equalities : $U(p_1m) = (Up_1)(Um) = r_1^{-1}q(Um) = r_1^{-1}q(Un) = (Up_1)(Un) = U(p_1n)$ imply $p_1m = p_1n$ and the existence of a morphism $s : P \to P_1$ satisfying $sp = p_1$. Then $(Us)r^{-1}q = (Us)(Up) = U(sp) = Up_1 = r_1^{-1}q$, thus $Us = r_1^{-1}r$ is an isomorphisme and hence so is s. The functor U thus

creates regular quotient objects. ∎

2.9.1. Theorem. "Beck's Theorem". If $\mathbb{B}$ is a locally simple category and $U : \mathbb{A} \to \mathbb{B}$ is a functor with target $\mathbb{B}$, the following assertions are equivalent :

(1) U is a monadic morphism [33] of locally simple categories.

(2) U is a monadic functor the monad of which is of finite rank and preserves finite products (1.11).

(3) U is a morphism of locally simple categories which reflects isomorphisms.

(4) $\mathbb{A}$ has generalised kernel pairs [35,21.5.11] and U has a left adjoint, preserves filtered colimits and creates regular quotient objects.

Proof : The implications (1) => (2) and (1) => (3) are immediate.

(2) => (1). The functor U is a monadic morphism of locally indecomposable categories (Proposition 1.11.0.). Since it is faithful, it reflects monomorphisms, thus is a morphism of locally simple categories (Proposition 2.8.2.).

(3) => (1). Since the functor U is exact (Proposition 12.8.1) it preserves coequalisers of equivalence relations, and thus is monadic by Beck's Theorem as modified by Duskin [35, Proposition 21.5.13].

(3) => (4). The functor U lifts regular quotient objects (Proposition 2.8.4), thus it created them (Proposition 2.9.0.1.).

(4) => (1). The functor U lifts isomorphisms, thus kernel pairs. It lifts quotients of equivalence relations since, for any equivalence relation $(m,n) : R \rightrightarrows X$ in $\mathbb{A}$, the coequaliser of (Um,Un) is of the form Uk where $k : X \to K$ is a regular epimorphism whose kernel pair is (m,n), since UK has as its kernel pair (Um,Un). It follows that k is the coequaliser of (m,n). U thus creates quotients of equivalence relations. The Duskin modification of Beck's Theorem [35,215.13] then implies that the functor U is monadic. The functor U lifts direct factors, since it lifts regular quotient objects and it reflects products. Moreover direct factors in $\mathbb{A}$ are strongly cocartesian for U, since U is monadic. By proposition 1.8.3., it follows that U is a morphism of locally indecomposable categories. By proposition 2.8.2., it follows that U is a morphism of locally simple categories. ∎

2.9.2. <u>Corollary</u>. <u>An algebraic functor $U : \mathbb{A}lg(\mathbb{T}') \to \mathbb{A}lg(\mathbb{T})$ between two categories of algebras in the sense of Lawvere, is a monadic morphism between locally simple categories if, and only if, the category $\mathbb{A}lg(\mathbb{T})$ is locally simple and U lifts regular quotient objects</u>.

<u>Proof</u> : Follows from Theorem 2.9.1. ■

2.9.3. <u>Corollary</u>. <u>If $\mathbb{A}$ is a locally simple category and A is an object of $\mathbb{A}$, the target functor $A/\mathbb{A} \to \mathbb{A}$ is a monadic morphism of locally simple categories</u>.

<u>Proof</u> : It is a monadic functor of finite rank, which has as a left adjoint the functor $A \coprod (-)$ which preserves finite products. ■

2.9.4. <u>Corollary</u>. <u>The composite of two monadic morphisms of locally simple categories, is monadic.</u>

2.9.5. <u>Examples</u>.

2.9.5.0. <u>$\mathbb{R}ngc\mathbb{R}eg\mathbb{S}t\mathbb{L}at \to \mathbb{R}ngc\mathbb{R}eg$</u>.
The forgetful functor U is algebraic. Let I be an ideal in a regular strongly lattice-ordered ring A. It is generated by idempotents. If $x \in I$, there is an idempotent, $e \in I$ satisfying $x = xe$. If $y \in A$ is such that $|y| \leqslant |x|$, then $|y-ye| = |y(1-e)| = |y||1-e| \leqslant |x||1-e| = x(1-e) = 0$, so $y = ye \in I$. The ideal I is thus absolutely convex. Moreover, if $x \in I$ and $y \in I$, there is an idempotent $e \in I$ satisfying $x = xe$ and $y = ye$, thus $x \vee y = ex \vee ey = e(x \vee y)$ since $e \geqslant 0$, and consequently, $x \vee y \in I$. The ideal I is thus closed under finite joins. It follows that the quotient ring A/I is strongly lattice-ordered and that the canonical projection $A \to A/I$ is a regular quotient object in $\mathbb{R}ngc\mathbb{R}eg\mathbb{S}t\mathbb{L}at$. The functor U thus lifts regular quotient objects. It is a monadic morphism of locally simple categories.

2.9.5.1. <u>$\mathbb{R}ngc\mathbb{B}a\mathbb{L}at \to \mathbb{R}ngc\mathbb{B}a$</u>.
The forgetful functor is algebraic and lifts regular quotient objects since any Baer ideal, I, of a Baer ring A, is generated by idempotents thus is closed under finite joins and consequently the canonical projection $A \to A/I$ is a regular quotient object in $\mathbb{R}ngc\mathbb{B}a\mathbb{L}at$. It is a monadic morphism of locally simple categories.

2.9.5.2. <u>$\mathbb{R}ngc\mathbb{D}if\mathbb{R}eg \to \mathbb{R}ngc\mathbb{R}eg$</u>. The forgetful functor is algebraic and lifts regular quotient objects since any ideal of a regular differential ring is a differential ideal.

2.9.5.3. <u>$\mathbb{A}$lgc$\mathbb{R}$eg(R) → $\mathbb{R}$ngc$\mathbb{R}$eg</u>.

If R ∈ $\mathbb{R}$ngc$\mathbb{R}$eg, the category R/$\mathbb{R}$ngcReg is isomorphic to the category $\mathbb{A}$lgc$\mathbb{R}$eg(R) of regular commutative algebras over R. The forgetful functor is thus a monadic morphism of locally simple categories. Starting from all locally simple categories of rings, one can construct in this way locally simple categories of algebras.

2.10. <u>LOCALLY SIMPLE VARIETIES</u>.

2.10.0. <u>Locally simple subcategories</u>.

Let $\mathbb{A}$ be a locally simple category. A <u>locally simple subcategory</u> of $\mathbb{A}$ is a subcategory of $\mathbb{A}$ such that the inclusion is a morphism of locally simple categories. By Proposition 2.8.2., this notion coincides with that of a locally indecomposable subcategory of $\mathbb{A}$. The results of Chapter 1 relevant to subcategories and to categories of fractions thus apply integrally. <u>Propositions 1.9.0. and 1.9.1. as well as Theorems 1.10.1 and 1.10.4 remain valid on substituting "simple" in place of "indecomposable"</u>. Taking account of Proposition 2.8.4, Proposition 1.9.0. can be completed in the following way :

2.10.0.0. <u>Proposition</u>. <u>A full subcategory $\mathbb{B}$ of $\mathbb{A}$ is a locally simple subcategory if and only if, it is a reflexive subcategory closed under filtered colimits and regular quotient objects</u>.

If $\mathbb{S}$ is a full subcategory of $\mathbb{S}$im $\mathbb{A}$, we will denote by $\mathbb{L}$oc$\mathbb{S}$, the full subcategory of $\mathbb{A}$, whose objects are <u>the objects of $\mathbb{A}$ locally in</u> $\mathbb{S}$ i.e. for which each simple component belongs to $\mathbb{S}$. Then $\mathbb{S}$im$\mathbb{L}$oc $\mathbb{S}$ = $\mathbb{S}$. Property (3) of Proposition 1.9.0. states precisely that $\mathbb{B}$ = $\mathbb{L}$oc$\mathbb{S}$im $\mathbb{B}$ for a locally simple full subcategory $\mathbb{B}$ of $\mathbb{A}$.

2.10.1. <u>Varieties of a locally simple category</u>.

As the category $\mathbb{A}$ is equivalent to an I-algebraic category (Corollary 2.4.1.), one can, following [9], describe the varieties of $\mathbb{A}$ in the following way : an <u>identity</u> of $\mathbb{A}$ is a pair, $(\omega,\mu) : X \rightrightarrows Y$, of morphisms of $\mathbb{A}$ whose source and target are finitely presentable projective objects. An object A of $\mathbb{A}$ <u>satisfies</u> the identity (ω,μ) if $\mathrm{Hom}_{\mathbb{A}}(\omega,A) = \mathrm{Hom}_{\mathbb{A}}(\mu, A)$ and it satisfies a set of identities if it satisfies each one of them. A variety of $\mathbb{A}$ is a full subcategory of $\mathbb{A}$ whose objects are those which satisfy a given set of identities.

2.10.1.0. Theorem. If $\mathbb{B}$ is a full subcategory of a locally simple category $\mathbb{A}$, the following assertions are equivalent :

(1) $\mathbb{B}$ is a variety of $\mathbb{A}$.

(2) $\mathbb{B}$ is a locally simple subcategory of $\mathbb{A}$, which is closed under subobjects.

(3) $\mathbb{B}$ is closed under products, subobjects and regular quotient objects.

(4) $\mathbb{B}$ is of the form, $\mathbb{L}oc\mathbb{S}$, where $\mathbb{S}$ is a full subcategory of $\mathbb{S}im\,\mathbb{A}$ which is closed under ultraproducts ans subobjects.

Proof : (1) => (2). Let $\mathbb{B}$ be the variety of $\mathbb{A}$ defined by a set, I, of identities of $\mathbb{A}$. If one denotes by Σ, the set of coequalisers of pairs $(\omega,\mu) \in I$, $\mathbb{B}$ is the category, $\mathbb{A}_{Loc\Sigma}$, of objects of $\mathbb{A}$ which are left closed for Σ, and so it is a locally simple subcategory. It is moreover immediate that $\mathbb{B}$ is closed under subobjects.

(2) => (3). Immediate.

(3) => (4). The simple components of an object A of $\mathbb{B}$ belong to $\mathbb{B}$, as they are regular quotient objects of A. Conversely if the simple components of an object A of $\mathbb{A}$, belong to $\mathbb{B}$, the object A belongs to $\mathbb{B}$ as it is a subobject of the product of its simple components. Thus one has $\mathbb{B} = \mathbb{L}oc\mathbb{S}im\mathbb{B}$. As the subcategory $\mathbb{B}$ is closed under subobjects, the subcategory $\mathbb{S}im\,\mathbb{B}$ is closed in $\mathbb{S}im\,\mathbb{A}$ under subobjects. As the two full subcategories, $\mathbb{B}$ and $\mathbb{S}im\,\mathbb{A}$ are closed in $\mathbb{A}$ under ultraproducts (Proposition 1.5.2.), the subcategory $\mathbb{S}im\,\mathbb{B}$ is closed in $\mathbb{S}im\,\mathbb{A}$ under ultraproducts.

(4) => (1). We will prove that $\mathbb{B}$ satisfies the conditions for the validity of Theorem 5.1.0., [9]. Let B be an object of $\mathbb{L}oc\mathbb{S}$ and $f : A \to B$ a monomorphism of $\mathbb{A}$. As the functor $\Delta : \mathbb{A} \to \mathbb{B}ool$ (1.3.5) preserves monomorphisms, the homomorphism $\Delta(f) : \Delta(A) \to \Delta(B)$ is a monomorphism and it follows that the mapping $Spec_{Max}(f) : Spec_{Max}(B) \to Spec_{Max}(A)$, is surjective. For each simple component $s : A \to S$ of A, there is therefore a simple component $f : B \to T$ of B and a monomorphism $g : S \to T$ satisfying $gs = tf$. However T is in $\mathbb{S}$, so S is in $\mathbb{S}$. Therefore A is in $\mathbb{L}oc\,\mathbb{S}$ and it follows that $\mathbb{L}oc\,\mathbb{S}$ is closed under subobjects. The category $\mathbb{L}oc\mathbb{S}$ is also closed under regular quotient objects, since if $q : A \to Q$ is

a regular quotient object of an object A of $\mathbb{L}\mathrm{oc}\mathbb{S}$, any simple component of Q is a simple component of A, hence belongs to $\mathbb{S}$. Let $(S_i)_{i\in I}$ be a family of objects of $\mathbb{S}$ and $q : \prod_{i\in I} S_i \to S$ a simple component of the product $\prod_{i\in I} S_i$. The set $U = \{I_0 \subset I : q$ factorises through the natural projection $\prod_{i\in I} S_i \to \prod_{i\in I_0} S_i\}$ is an ultrafilter on I. The ultraproduct S_U of the family $(S_i)_{i\in I}$ relative to the ultrafilter U is an object of $\mathbb{S}$ and the natural projection $q_U ; \prod_{i\in I} S_i \to S_U$ factorises the morphism q by a morphism $f : S_U \to S$, which is necessarily an isomorphism. The object S thus belongs to $\mathbb{S}$ and the product $\prod_{i\in I} S_i$ to $\mathbb{L}\mathrm{oc}\mathbb{S}$. Let $(A_i)_{i\in I}$ be a family of objects of $\mathbb{L}\mathrm{oc}\,\mathbb{S}$. Since each object A_i is a subobject of the product of its simple components, the product $\prod_{i\in I} A_i$ is itself a subobject of a product of objects of $\mathbb{S}$, thus belongs to $\mathbb{L}\mathrm{oc}\mathbb{S}$. It follows that $\mathbb{L}\mathrm{oc}\mathbb{S}$ is closed under products. Let $D : \mathbb{I} \to \mathbb{S}\mathrm{im}\,\mathbb{B}$ be a small filtered diagram of simple objects of $\mathbb{B}$. Let I be the set of objects of $\mathbb{I}$ preordered by the relation $i \leqslant j \Leftrightarrow \exists f : i \to j$ in $\mathbb{I}$. It is a filtered preordered set and the canonical functor $Q : \mathbb{I} \longrightarrow I$ is final. [33, p.283]. The diagram D factorises through Q because, for two morphisms $f,g : i \rightrightarrows j$ of $\mathbb{I}$, there is a morphism $h : j \to k$ of $\mathbb{I}$ satisfying $hf = hg$ and hence $(Dh)(Df) = (Dh)(Dg)$, which implies $Df = Dg$, since Dh is a monomorphism. If one still denotes by $D : I \to \mathbb{S}\mathrm{im}\,\mathbb{B}$ the functor thus obtained, the two diagrams, D, have the same colimit in $\mathbb{A}$. For each $i \in I$, set $K_i = \{j \in I : i \leqslant j\}$. The set $\{K_i\}_{i\in I}$ is a filter base on I. Let F be an ultrafilter on I finer than $\{K_i\}_{i\in I}$. The assignment $i \mapsto K_i$ defines a functor $K : I \to F^{op}$. Denote by $\Pi : F^{op} \to \mathbb{B}$ the functor defined by $\Pi J = \prod_{i\in J} D_i$ and the canonical projections. One defines a natural transformation $\alpha : D \to \Pi K$ by $\alpha_i = (D(f_{ji}))_{j\in K_i} : D_i \to \prod_{j\in K_i} D_j$ where for $i \leqslant j$, $f_{ji} : i \to j$ is some morphism in $\mathbb{I}$. This natural transformation is a monomorphism since α_i is a split monomorphism. It induces a monomorphism, $\varinjlim D \to \varinjlim \Pi$. However this latter colimit is the ultraproduct of the family of simple objects, $(D_i)_{i\in I}$, relative to the ultrafilter F. It is thus a simple object and it follows that the colimit of D is also a simple object. Let $(A_i)_{i\in\mathbb{I}}$ be a filtered diagram of objects of $\mathbb{L}\mathrm{oc}\,\mathbb{S}$ with colimit $(\alpha_i : A_i \to A)_{i\in\mathbb{I}}$. Let $s : A \to S$ be a simple component. For each $i \in \mathbb{I}$, set $\mathrm{Spec}_{\mathrm{Max}}(\alpha_i)(s) = s_i : A_i \to S_i$ and denote $\beta_i : S_i \to S$ the morphism satisfying $\beta_i s_i = s\alpha_i$. For each $f : i \to j$ in $\mathbb{I}$, let $S_f : S_i \to S_j$ be the morphism satisfying $s_f s_i = s_j A_f$. One thus obtains a filtered diagram $(S_i)_{i\in\mathbb{I}}$ whose colimit is $(\beta_i : S_i \to S)_{i\in\mathbb{I}}$. Since the objects S_i belong to $\mathbb{S}$, the object S belongs to $\mathbb{S}$. This proves that A belongs to $\mathbb{L}oc\,\mathbb{S}$ and consequently that $\mathbb{L}oc\,\mathbb{S}$ is closed under filtered colimits. ■

2.10.1.1. Proposition. Any variety of a locally simple category is generated by its simple objects.

Proof : Let $\mathbb{B}$ a variety of $\mathbb{A}$. Each object of $\mathbb{B}$ is a subobject of the product of its simple components, thus is a subobject of a product of objects from $\mathbb{S}im\,\mathbb{B}$. The variety of $\mathbb{A}$ generated by $\mathbb{S}im\,\mathbb{B}$ contains all subobjects of products of objects from $\mathbb{S}im\,\mathbb{B}$, thus contains $\mathbb{B}$, and hence is $\mathbb{B}$. ■

2.10.2. Varieties of locally closed objects.

Let Γ be a set of finite families of regular epimorphisms in $\mathbb{A}$, having the same source and whose sources and targets are finitely presentable. Following [13, 3.1.], an object A of $\mathbb{A}$ is left closed for Γ if, for any family $(\gamma_i : X \to X_i)_{i\in[1,n]}$ of Γ, any morphism $f : X \to A$ factorises through one of the morphisms, γ_i.

2.10.2.0. Definition. An object of $\mathbb{A}$ is locally left closed for Γ if its simple components are left closed for Γ.

These objects generate the full subcategory $\mathbb{A}_{loc\Gamma}$ of $\mathbb{A}$.

2.10.2.1. Theorem. The category $\mathbb{A}_{loc\Gamma}$ of objects of $\mathbb{A}$ which are locally left closed for Γ is a variety of $\mathbb{A}$, whose reflector sends the families of morphisms of Γ onto counions of quotient objects, and such that, any morphism of locally simple categories $U : \mathbb{B} \to \mathbb{A}$ whose left adjoint sends the families of morphisms of Γ onto counions of quotient objects, factorises in a unique way through $\mathbb{A}_{loc\Gamma}$ by a morphism of locally simple categories.

Proof : Let $(\gamma_i : X \to X_i)_{i\in[1,n]}$ be a family of morphisms in Γ. Since the regular epimorphisms γ_i have, for their sources and their targets, finitely presentable objects, they are finitely generated and thus are direct factors. For each $i \in [1,n]$, set

$\delta_i = \gamma_i \wedge \gamma'_{i-1} \wedge \ldots \wedge \gamma'_1$. The direct factors δ_i are pairwise disjoint and satisfy $\delta_1 \vee \ldots \vee \delta_n = \gamma_1 \vee \ldots \vee \gamma_n$. Moreover, $\delta_i \leqslant \gamma_i$ i.e. δ_i factors through γ_i. Let S be a simple object of $\mathbb{A}$. If it is left closed for the family $(\delta_i)_{i\in[1,n]}$, it is also for the family $(\gamma_i)_{i\in[1,n]}$. Conversely, suppose it is left closed for $(\gamma_i)_{i\in[1,n]}$. Let $f : X \to S$ and let j be the smallest integer of $[1,n]$ such that f factorises through γ_j. Then f does not factorise through $\gamma_1,\ldots,\gamma_{j-1}$, thus factorises through $\gamma'_i,\ldots,\gamma'_{j-1}$ and hence through $\gamma_j \wedge \gamma'_{j-1} \wedge \ldots \wedge \gamma'_1 = \delta_j$. Thus S is left closed for the family $(\delta_i)_{i\in[1,n]}$. As this family is composed of pairwise disjoint direct factors, the object S is left closed for (δ_i) if, and only if, it is strictly left closed for (δ_i) [13]. As a result an object A of $\mathbb{A}$ is locally left closed for $(\ _i)$ if, and only if, it is locally strictly left closed for (δ_i) and consequently, if one denoted by Δ, the set of families (δ_i) associated to the families (γ_i) of Γ, one has $\mathbb{A}_{loc\Gamma} = \mathbb{A}_{loc(\Delta)}$. By theorem 1.10.4., it follows that $\mathbb{A}_{loc\Gamma}$ is a locally simple subcategory of $\mathbb{A}$ and that the inclusion functor is a universal morphism of locally simple categories such that the reflector sends the families of Δ onto products. However for a morphism of locally simple categories $U : \mathbb{B} \to \mathbb{A}$, the left adjoint $F : \mathbb{A} \to \mathbb{B}$ preserves direct factors, their cointersections, their complements and thus their counions; it thus sends the families of Δ to products if, and only if, it sends those of Γ to counions of regular quotient objects. It is, moreover, immediate that $\mathbb{A}_{loc\Gamma}$ is closed under subobjects since the morphisms in the families of Γ are regular epimorphisms, consequently $\mathbb{A}_{loc\Gamma}$ is a variety of $\mathbb{A}$, (Proposition 2.10.1.0). ■

2.10.3. Examples. The categories $\mathbb{R}$ngc$\mathbb{R}$eg, $\mathbb{R}$ngc$\mathbb{R}$eg(p), p-$\mathbb{R}$ngc, p^k-$\mathbb{R}$ngc, $\mathbb{R}$ngc$\mathbb{R}$eg$\mathbb{F}$orm$\mathbb{R}$l and $\mathbb{B}$ool are varieties of $\mathbb{R}$ng$\mathbb{S}$tr$\mathbb{R}$eg. The category $\mathbb{R}$ngc$\mathbb{D}$if$\mathbb{R}$eg$\mathbb{P}$erf(p) is a locally simple subcategory of $\mathbb{R}$ngc$\mathbb{D}$if$\mathbb{R}$eg and the category $\mathbb{R}$ngc$\mathbb{D}$if$\mathbb{R}$eg(p) is a variety of it. $\mathbb{R}$ngc$\mathbb{B}$a$\mathbb{S}$tr$\mathbb{L}$at is a variety of $\mathbb{R}$ngc$\mathbb{B}$a$\mathbb{L}$at , and Rngc$\mathbb{E}$ is a variety of Rngc$\mathbb{E}$.

2.11. LOCALLY SIMPLE ALGEBRAIC THEORIES.

2.11.0. Following Kennison [26, 4], a comparison on a set E is a quaternary operation $C : E^4 \to E$ satisfying the five axioms :

(C1) $C(a,a,x,y) = x$

(C2) $C(a,b,x,x) = x$

(C3) $C(a,b,x,y) = C(b,a,x,y)$

(C4) $C(a,b,a,b) = b$

(C5) $C(a,b,C(x_1,x_2,x_3,x_4), C(y_1,y_2,y_3,y_4)) =$

$C(C(a,b,x_1,y_1), C(a,b,x_2,y_2), C(a,b,x_3,y_3), C(a,b,x_4,y_4))$.

For example the direct comparison on a set E is defined by :

$C(a,b,x,y) = x$ for $a = b$ and

$C(a,b,x,y) = y$ for $a \neq b$.

2.11.1. Definition. A locally simple algebraic theory is an algebraic theory $\mathbb{T}$ in the sense of Lawvere [30] having at least two constants, denoted 0 and 1, and a quaternary comparison operation C, satisfying the following axioms :

(C6) for any n-ary operation ω of $\mathbb{T}$,

$C(a,b,\omega(x_1,\ldots,x_n), \omega(y_1,\ldots,y_n)) =$

$\omega(C(a,b,x_1,y_1),\ldots,C(a,b,x_n,y_n))$

(C7) $C(0,1,x,0) = 0$.

2.11.2. Idempotents.

If $\mathbb{T}$ is a locally simple algebraic theory and E is a $\mathbb{T}$-algebra, an element e of E is said to be idempotent if it satisfies $e = C(0,e,0,1)$. The compatible equivalence relation on E generated by the pair $(0,e)$ is called the relation of equivalence modulo e. It will be denoted $x \sim y \pmod{(e)}$, and the quotient of E by this relation will be denoted $E/(e)$. By Kennison [30, lemma 4.6], one has :
$x \sim y(\mathrm{mod}(e)) \iff C(0,e,x,y) = y$.

2.11.3. Proposition. The mapping $e \mapsto E/(e)$ is a bijection between the set of idempotents of E and the set of direct factors of E.

Proof : For each idempotent e of E, set $e' = c(0,e,1,0)$

and let us show that $(E \to E/(e), E \to E/(e'))$ is a product. It is easy to see that e' is an idempotent. If $x,y \in E$ are such that $x \sim y(\mathrm{mod}(e))$ and $x \sim y(\mathrm{mod}(e'))$ then $y = C(0,e,x,y) = C(0,C(0,e,1,0),x,v) = C(0,e,C(0,1,x,y), C(0,0,x,y)) = C(0,e,y,x) = x$. For each pair, (x,y), of elements of E, the element $C(0,e,x,y)$ is equivalent to x modulo (e) since $C(0,e\ C(0,e,x,y),x) = C(0,e,x,x) = x$ and also equivalent to y modulo (e'), since $C(0,e', C(0,e,x,y),y) = C(0,C(0,e,1,0), C(0,e,x,y),y) = C(0,e,C(0,1,x,y), C(0,0,y,y)) = C(0,e,y,y) = y$. The pair $(E \to E/(e), E \to E/(e'))$ is thus a product as claimed, and $E \to E/(e)$ is a direct factor of E. The mapping $e \mapsto E/(e)$ is injective since, if e and f are two idempotents such that $E/(e) = E/(f)$, then $f \sim 0(\mathrm{mod}(e))$ and $e \sim 0(\mathrm{mod}(f))$, thus $f = C(0,e,0,f) = C(0,e,0,C(0,f,0,1)) = C(0,f,C(0,e,0,0), C(0,e,0,1)) = C(0,f,0,e) = e$. It is surjective since, if $(p : E \to F,\ q : E \to G)$ is the product of the $\mathbb{T}$-algebras F and G, the morphism p is isomorphic to the quotient of E modulo the idempotent $(0,1)$. ■

2.11.4. Theorem. If $\mathbb{T}$ is a locally simple algebraic theory, the category $\mathbb{A}\mathrm{lg}(\mathbb{T})$ is locally simple. Its simple objects are the $\mathbb{T}$-algebras in which the comparison is direct and $0 \neq 1$.

Proof : First we will study the algebraic theory $\mathbb{T}_0$, generated by two constants 0 and 1 and a comparison operator C satisfying $C(0,1,x,0) = 0$. The initial $\mathbb{T}_0$-algebra is $\{0,1\}$ equipped with the direct comparison. Its square is a finite set thus is a finitely presentable $\mathbb{T}_0$-algebra. The final $\mathbb{T}_0$-algebra is $\{0\}$ and thus is a strict final object. Since $\mathbb{T}_0$-algebras are not empty, products of two $\mathbb{T}_0$-algebras are codisjoint. Let us show they are couniversal. Let $(p : E\times F \to E,\ q : E\times F \to F)$ be the product of two $\mathbb{T}_0$-algebras E and F and let $f : E\times F \to G$ be a $\mathbb{T}_0$-homomorphism. The element $e = f(0,1)$ is idempotent and $e' = f(1,0)$. Since p (resp. q) is isomorphic to the quotient of $E\times F$ by $(0,1)$ (resp. $(1,0)$), the direct image of p (resp. q) by f is isomorphic to the quotient of G modulo (e) (resp. (e')). The direct image of (p,q) by f is thus the product $(G \to G/(e),\ G \to G/(e'))$. As a result, the category $\mathbb{A}\mathrm{lg}(\mathbb{T}_0)$ is locally indecomposable. Let E be a $\mathbb{T}_0$-algebra and $a,b \in E$. The element $e = C(a,b,0,1)$ is idempotent since $C(0,C(a,b,0,1),0,1) = C(a,b,C(0,0,0,1), C(0,1,0,1)) = C(a,b,0,1)$. The relation of equivalence modulo (e) is the equivalence relation generated by the pair (a,b) since the equalities $C(0,e,x,y) = C(0,C(a,b,0,1),x,y) = C(a,b,C(0,0,x,y), C(0,1,x,y)) = C(a,b,x,y)$ imply

$x \sim y(\mathrm{mod}(e)) \iff C(0,e,x,y) = y \iff C(a,b,x,y) = y$ [26, lemma 4.6]. The quotient of E by the equivalence relation generated by (a,b) is thus a direct factor of E. This proves that any indecomposable object of $\mathbb{A}\mathrm{lg}(\mathbb{T}_0)$ is a simple object and it follows that $\mathbb{A}\mathrm{lg}(\mathbb{T}_0)$ is a locally simple category (Proposition 2.1.2).

The final $\mathbb{T}_0$-algebra $\{0\}$ is characterised by the property $0 = 1$, since that implies : $x = C(0,0,x,0) = C(0,1,x,0) = 0$. If E is a simple $\mathbb{T}_0$-algebra then $E \neq \{0\}$, thus $0 \neq 1$ and if a,b are two distinct elements of E, the equivalence on E generated by the pair (a,b) is the coarse equivalence relation and it follows that, for any $x,y \in E$, one has $C(a,b,x,y) = y$, which proves the comparison on E is direct. Conversely, if E is a $\mathbb{T}_0$-algebra in which the comparison is direct and $0 \neq 1$, then $E \neq 0$ and any non null idempotent element e of E satisfies $e = C(0,e,0,1) = 1$, thus E is simple.

Next let us study a locally simple algebraic theory, $\mathbb{T}$. The forgetful functor $U : \mathbb{A}\mathrm{lg}(\mathbb{T}) \to \mathbb{A}\mathrm{lg}(\mathbb{T}_0)$ lifts direct factors because, if A is a $\mathbb{T}$-algebra and e is an idempotent of A, the relation of equivalence modulo (e) is compatible with the $\mathbb{T}$-algebra structure and hence the quotient $E \to E/(e)$ lifts to $\mathbb{A}\mathrm{lg}(\mathbb{T})$. The functor U thus lifts regular quotient objects and is thus a morphism of locally simple categories (Corollary 2.9.2.) and it preserves and reflects the simple objects. ∎

2.11.5. <u>Theorem</u>. <u>Any locally simple algebraic category is equivalent to a category of algebras for a locally simple algebraic theory</u>.

<u>Proof</u> : Let $\mathbb{A}$ be a locally simple algebraic category. $\mathbb{A}$ has a projective finitely presentable proper generating object G_0. The object $G = Z^2 \times G_0$ is also a projective finitely presentable proper generator (cf. Proof. 2.4.0 (c)). The functor $\mathrm{Hom}_{\mathbb{A}}(G,-) : \mathbb{A} \to \mathbb{E}\mathrm{ns}$ is algebraic. It generates the algebraic theory $\mathbb{T}$, which is equivalent to the dual of the full subcategory of $\mathbb{A}$ whose objects are of the form nG, sum of n copies of the object G. Let us identify these two categories. The two projections, $G = Z \times Z \times G_0 \rightrightarrows Z = 0G$, define the two constants $0,1$ of $\mathbb{T}$. Let us denote by $\nabla : 2G \to G$, the codiagonal of G and $\nabla' = 2G \to G'$, its complement. Then $(\nabla,\nabla') : 2G \to G \times G'$ is an isomorphism. Set : $\Gamma = \nabla \coprod 1_G \coprod 1_G : 4G \to 3G$ and $\Gamma' = \nabla' \coprod 1_G \coprod 1_G : 4G \to G' \coprod 2G$. Then

$(\Gamma,\Gamma') : 4G \to 3G\times(G' \coprod 2G)$ is an isomorphism. Let $C : G \to 4G$ be the morphism such that $\Gamma C : G \to 3G$ is the second canonical injection and $\Gamma' C : G \to G' \coprod G \coprod G$ is the third canonical injection. This morphism defines a quaternary operation $C : 4 \to 1$ of $\mathbb{T}$. The functor $\mathrm{Hom}_{\mathbb{A}}(G,-) : \mathbb{A} \to \mathbb{E}\mathrm{ns}$ lifts to an equivalence of categories $K : \mathbb{A} \to \mathbb{A}\mathrm{lg}(\mathbb{T})$ defined for an object A of $\mathbb{A}$ by $KA(\omega) = \mathrm{Hom}_{\mathbb{A}}(\omega,A)$ for an n-ary operation ω of $\mathbb{T}$. Consider a simple object S of $\mathbb{A}$. If $a,x,y \in \mathrm{Hom}_{\mathbb{A}}(G,S)$, the morphism $<a,a.x,y> : 4G \to S$ factorises through $\Gamma : 4G \to 3G$ in the form $<a,a,x,y> = <a,x,y>\Gamma$ and consequently $KS(C)\ (<a,a,x,y>) = <a,a,x,y>C = <a,x,y>\Gamma C = x$. If $a,b,x,y \in \mathrm{Hom}_{\mathbb{A}}(G,S)$ are such that $a \neq b$, the morphism $<a,b,x,y> : 4G \to S$ does not factorise through Γ, hence must factorise through Γ' in the form $<a,b,x,y> = <d,x,y>\Gamma'$ and consequently $KS(C)(<a,b,x,y>) = <a,b,x,y>C = <d,x,y>\Gamma' C = y$. The operation $KS(C) : \mathrm{Hom}_{\mathbb{A}}(4G,S) \to \mathrm{Hom}_{\mathbb{A}}(G,S)$ is thus identifiable as the direct comparison on $\mathrm{Hom}_{\mathbb{A}}(G,S)$. Moreover it satisfies $KS(C)(<0,1,x,0>) = 0$ since $0 \neq 1$ in all simple objects of $\mathbb{A}$. Consider one of the axioms (C1),...,(C7) of locally simple algebraic theories. It is of the form $\omega(x_1,\dots,x_n) = \mu(x_1,\dots,x_n)$ where ω,μ are two n-ary operations of $\mathbb{T}$. For each simple object S of $\mathbb{A}$, the $\mathbb{T}$-algebra, $\mathrm{Hom}_{\mathbb{A}}(G,S)$ satisfies this axiom, since the operation defined by C is the direct comparison, so $KS(\mu) = KS(\omega)$. The inclusion functor $J : \mathbb{S}\mathrm{im}\,\mathbb{A} \to \mathbb{A}$ is cogenerating and so, it follows that the functor $J' : \mathbb{A}^{op} \to [\mathbb{S}\mathrm{im}\,\mathbb{A},\mathbb{E}\mathrm{ns}]$ defined by $J'(\cdot) = \mathrm{Hom}_{\mathbb{A}}(\cdot,J(-))$ is faithful. However the two morphisms $\omega,\mu : G \to nG$ satisfy $J'(\omega)_S = \mathrm{Hom}_{\mathbb{A}}(\omega,J(S)) = KS(\omega) = KS(\mu) = \mathrm{Hom}_A(\mu,J(S)) = J'(\mu)_S$ thus $J'(\omega) = J'(\mu)$ and $\omega = \mu$. The axioms C(1)...(C7) are thus satisfied in $\mathbb{T}$, which proves that the theory $\mathbb{T}$ is locally simple. ∎

2.12. TOTALLY SEPARATED BOOLEAN SHEAVES OF ALGEBRAS.

Let $\mathbb{T}$ be an algebraic theory in the sense of Lawvere, having two distinguished constants, denoted 0 and 1.

2.12.0. Definition. A sheaf of $\mathbb{T}$-algebras $F : \Omega(X)^{op} \to \mathbb{A}\mathrm{lg}(\mathbb{T})$ is totally separated if the topological space X and the étale space associated to F are both separated spaces and if $0 \neq 1$ in each stalk.

2.12.1. Proposition. If X is a Boolean topological space, a sheaf of $\mathbb{T}$-algebras on X is totally separated if, and only if, for each pair (s,t) of continuous global sections, the set

<u>$E(s,t) = \{x \in X : s(x) = t(x)\}$ is an open-closed set of X and $E(0,1) = \emptyset$.</u>

<u>Proof</u> : Let $F : \Omega(X)^{op} \to \mathbb{A}lg(\mathbb{T})$ be a totally separated sheaf of $\mathbb{T}$-algebras. One knows that the sets $E(s,t)$ are open in X. If F is totally separated, these sets are closed since they are the equalisers of two continuous maps with values in a separated space. Conversely, suppose that the sets $E(s,t)$ are closed. Let $x \in X$ and α,β two distinct elements of F_x. There exists an open-closed neighbourhood u of x and two sections s, t of F defined on u such that $s(x) = \alpha$ and $t(x) = \beta$. Define two global sections, $\bar{s},\bar{t}$, of F in such a way that their restrictions to u are respectively s,t and their restrictions to $X-u$ have value 0. Then $v = \{y \in X : \bar{s}(y) \neq \bar{t}(y)\} = X - E(\bar{s},\bar{t})$ is an open-closed neighbourhood of x on which the restrictions of $\bar{s},\bar{t}$ define two disjoint open sets of the associated étale space of F, containing respectively α,β. The associated étale space of F is thus separated and F is totally separated. ■

2.12.2. <u>Proposition. If $\mathbb{T}$ is a locally simple algebraic theory, every Boolean sheaf of simple $\mathbb{T}$-algebras is totally separated</u>.

<u>Proof</u> : Let X be a Boolean topological space and $F : \Omega(X)^{op} \to \mathbb{A}lg(\mathbb{T})$ a sheaf of $\mathbb{T}$-algebras on X whose stalks F_x are simple $\mathbb{T}$-algebras i.e. $\mathbb{T}$-algebras in which the comparison is direct and $0 \neq 1$. If s,t are two global sections of F, the set $\{x \in X : s(x) \neq t(x)\} = \{x \in X : C(s(x), t(x),1,0) = 0\} = \{x \in X : C(s,t,1,0)(x) = 0(x)\}$ is an open set of X, thus the set $E(s,t)$ is closed in X and thus is an open-closed set of X. ■

The category $\mathbb{S}hvBoolTsep\mathbb{A}lg(\mathbb{T})$ <u>of totally separated Boolean sheaves of $\mathbb{T}$-algebras</u> has, as object, the pairs (X,F), where X is a Boolean space and F is a totally separated sheaf of $\mathbb{T}$-algebras F, with base X, and for morphisms, it has those pairs (f,α) formed from a continuous map $f : Y \to X$ and a natural transformation $\alpha : F \to Gf^*$ such that the induced morphisms α_y on the stalks, are injective. If $\mathbb{M}$ is a full subcategory of $\mathbb{A}lg(\mathbb{T})$, the <u>category</u> $\mathbb{S}hvBoolTsep\,\mathbb{M}$ is the full subcategory of $\mathbb{S}hvBoolTsep\mathbb{A}lg(\mathbb{T})$, whose objects are those pairs (X,F) in which the stalks of F are in $\mathbb{M}$.

2.12.3. <u>Theorem</u>. <u>For any algebraic theory, $\mathbb{T}$, having two distinguished constants 0 and 1, the category $\mathbb{S}hvBoolTsep\mathbb{A}lg(\mathbb{T})$ of</u>

totally separated Boolean sheaves of $\mathbb{T}$-algebras is locally simple and is equivalent to the category, $\mathbb{A}lg(\mathbb{CT})$, of $\mathbb{CT}$-algebras where $\mathbb{CT}$ is the locally simple algebraic theory obtained on adjoining to $\mathbb{T}$ a comparison operation. Its simple objects can be identified with the non null $\mathbb{T}$-algebras.

Proof : The category $\mathbb{A}lg(\mathbb{CT})$ is locally simple and thus is equivalent to the category, $\mathfrak{S}hvBool\mathfrak{S}im\mathbb{A}lg(\mathbb{CT})$, (Theorem 2.7.0). The forgetful functor $U : \mathbb{A}lg(\mathbb{CT}) \to \mathbb{A}lg(\mathbb{T})$ induces, following Proposition 2.12.2, a functor $U_* : \mathfrak{S}hvBool\mathfrak{S}im\mathbb{A}lg(\mathbb{CT}) \to \mathfrak{S}hvBoolTsep\mathbb{A}lg(\mathbb{T})$ defined by $U_*(X,F) = (X,UF)$ and $U_*(f,\alpha) = (f,U\alpha)$. Let us show that U_* is an equivalence of categories. This functor is faithful since U is faithful. Let (X,F), (Y,G) be two objects of $\mathfrak{S}hvBool\mathfrak{S}im\mathbb{A}lg(\mathbb{CT})$ and $(f,\alpha) : (X,F) \to (Y,G)$ a morphism of $\mathfrak{S}hvBoolTsep\mathbb{A}lg(\mathbb{T})$. Let u be an open set of X and $v = f^{-1}(u)$. For $x \in u$ and $y \in v$, denote by $\rho_x^u : Fu \to F_x$ and $\rho_y^v : Gv \to G_y$ the restriction morphisms of F and G. The morphisms α_y are monomorphisms and thus are $\mathbb{CT}$-homomorphisms. The product morphism $\prod_{y\in v} \alpha_y : \prod_{y\in v} F_{f(y)} \to \prod_{y\in v} G_y$ is thus a $\mathbb{CT}$-homomorphism, as is the composed morphism $(\prod_{y\in v} \alpha_y)(\rho^u_{f(y)})_{y\in v} : Fu \longrightarrow \prod_{y\in v} G_y$. However this latter morphism is equal to $(\rho_y^v)_{y\in v}\, \alpha_u : Fu \to \prod_{y\in v} G_y$. Since $(\rho_y^v)_{y\in v} : Gv \longrightarrow \prod_{y\in v} G_y$ is an injective $\mathbb{CT}$-homomorphism, it follows that α_u is a $\mathbb{CT}$-homomorphism. The pair (f,α) is thus a morphism of $\mathfrak{S}hvBool\mathfrak{S}im\mathbb{A}lg(\mathbb{CT})$ which proves that the functor U_* is full. Let (X,F) be an object of $\mathfrak{S}hvBoolTsep\mathbb{A}lg(\mathbb{T})$, u an open set of X and $a,b,s,t \in Fu$. The set $E(a,b) = \{x\in u : a(x) = b(x)\}$ is an open-closed set of u. For each $x\in E(a,b)$, one has $C(a(x),b(x),s(x),t(x)) = s(x)$ and, for $x\in u - E(a,b)$, one has $C(a(x),b(x),s(x),t(x)) = t(x)$. Thus $C(a,b,s,t)$ is a continuous section of F on u. It follows that Fu is a sub-$\mathbb{T}$-algebra of $\prod_{x\in u} F_x$, closed under the comparison operation, thus a $\mathbb{CT}$-algebra. Since the restriction morphisms of F are $\mathbb{CT}$-homomorphisms, (X,F) is the image by U_* of an object of $\mathfrak{S}hvBool\mathfrak{S}im\mathbb{A}lg(\mathbb{CT})$. ■

2.12.4. Theorem. If $\mathbb{M}$ is a full subcategory of $\mathbb{A}lg(\mathbb{T})$ closed under ultraproducts and subobjects, the category $\mathfrak{S}hvBoolTsep\,\mathbb{M}$ is a locally simple algebraic category. The categories of this form are exactly the varieties of $\mathfrak{S}hvBoolTsep\mathbb{A}lg(\mathbb{T})$.

Proof : The category of simple objects of $\mathbb{S}hvBoolTsep\mathbb{A}lg(\mathbb{T})$ is equivalent to the category $\mathbb{A}lgMono(\mathbb{T})$ having as its objects, the non-null $\mathbb{T}$-algebras and for its morphisms the monomorphic $\mathbb{T}$-homomorphisms. By theorem 2.10.1.0, the varieties of $\mathbb{S}hvBoolTsep\mathbb{A}lg(\mathbb{T})$ are the full subcategories of the form $\mathbb{L}oc\,\mathbb{M}$ where $\mathbb{M}$ is a full subcategory of $\mathbb{A}lgMono(\mathbb{T})$ closed under ultraproducts and subobjects. However the categories of the form $\mathbb{L}oc\,\mathbb{M}$ are precisely the categories of the form $\mathbb{S}hvBoolTsep\,\mathbb{M}$. ∎

2.12.5. Theorem. Locally simple algebraic categories are precisely the categories equivalent to categories of the form $\mathbb{S}hBoolTsep\mathbb{M}$ where $\mathbb{M}$ is a full subcategory, closed under ultraproducts and subobjects, of a category of $\mathbb{T}$-algebras for an algebraic theory having at least two constants.

Proof : It follows immediately from Theorems 2.7.0, 2.11.4, 2.12.4., and from Proposition 2.12.2. ∎

2.12.6. Remarks. The equivalence of the two categories $\mathbb{S}hvBoolTsep\mathbb{A}lg(\mathbb{T})$ and $\mathbb{A}lg(\mathbb{C}\mathbb{T})$ and the fact that the categories $\mathbb{S}hvBoolTsep\,\mathbb{M}$ are the varieties of $\mathbb{S}hvBoolTsep\mathbb{A}lg(\mathbb{T})$ are also consequences of the work of Kennison [26] : in fact, the category $\mathbb{S}hvBoolTsep\mathbb{A}lg(\mathbb{T})$ is a variety of the category $\mathbb{C}anSepShf(\mathbb{T})^{op}$, dual to the category of canonical separated sheaves of $\mathbb{T}$-algebras defined by Kennison. Also the category $\mathbb{A}lg(\mathbb{C}\mathbb{T})$ is a variety of the category $\mathbb{A}lg(\mathbb{C} \# \mathbb{T})$ defined by Kennison. The results then follow, in particular, from Theorem 5.1. [26], which states the equivalence $\mathbb{C}anSepShf(\mathbb{T})^{op} \sim \mathbb{A}lg(\mathbb{C} \# \mathbb{T})$ and from Theorem 6.7 [26] describing the Birkhoff subcategories of $\mathbb{A}lg(\mathbb{C} \# \mathbb{T})$.

2.13. LOCALLY SIMPLE CATEGORIES OF COALGEBRAS FOR A COMONAD.

We will use the notion of a comonad of finite rank preserving finite products, defined in 1.12.0.

2.13.0. Proposition. If $\mathbb{A}$ is a locally simple category and $\mathbb{L}$ is a comonad of finite rank on $\mathbb{A}$ which preserves finite products, then the category $\mathbb{A}^{\mathbb{L}}$ of $\mathbb{L}$-coalgebras is locally simple.

Proof : The category $\mathbb{A}^{\mathbb{L}}$ is locally indecomposable (Proposition 1.12.2.). Let (A,a) be a finitely presentable object of $\mathbb{A}^{\mathbb{L}}$. The sum of (A,a) with itself is of the form $(A \coprod A,\alpha)$ and the codiagonal of (A,a) is $\nabla : (A \coprod A,\alpha) \to (A, a)$ where $\nabla : A \coprod A \to A$ is the codiagonal of A. Since the object A is finitely presentable in $\mathbb{A}$, the codiagonal of A is a complemented quotient object (Proposition 2.1.2.) and forms with its complement $\nabla' : A \coprod A \to A'$, a product whose image by L is a product $(L\nabla,L\nabla')$. Since finite products are couniversal in $\mathbb{A}$ and $L\nabla$ is the direct image of ∇ by α, $L\nabla'$ is the direct image of ∇' by α, and there exists a morphism $a' : A' \to LA'$ satisfying $a'\nabla' = (L\nabla')\alpha$. Thus one obtains an $\mathbb{L}$-coalgebra (A',a') and a morphism of $\mathbb{L}$-coalgebras $\nabla' : (A \coprod A,\alpha) \to (A',a')$ forming with $\nabla : (A \coprod A,\alpha) \to (A,a)$, a product in $\mathbb{A}^{\mathbb{L}}$. The object (A,a) is thus codecidable. The category $\mathbb{A}^{\mathbb{L}}$ is thus locally simple (Proposition 2.1.2). ■

2.13.1. Examples.

2.13.1.1. The category $\mathbb{A}$lgcAlgReg(k) of regular commutative k-algebras which are algebraic over a commutative field k, is comonadic for an idempotent comonad of finite rank which preserves finite products, on the category $\mathbb{A}$lgcReg(k) of regular commutative k-algebras.

The category $\mathbb{A}$lgcAlgReg(k) is the full subcategory of $\mathbb{A}$lgcReg(k) whose objects are the regular algebraic k-algebras. One shows, as in 1.12.3, that one can define an idempotent comonad of finite rank, $\mathbb{L}$, which preserves finite products, on $\mathbb{A}$lgcReg(k). The functor L associates to each regular k-algebra A, the algebraic closure of k in A, i.e. the set of elements of A which are algebraic over k. The category $\mathbb{A}\text{lgcReg}(k)^{\mathbb{L}}$ is then equivalent to $\mathbb{A}$lgcAlgReg(k). Since the category, $\mathbb{A}$lgcReg(k), is locally simple, the category $\mathbb{A}$lgcAlgReg(k) is locally simple. ■

2.13.1.2. The category $\mathbb{A}$lgcAlgSep(k) of separable algebraic commutative k-algebras over a commutative field k is comonadic for a comonad of finite rank which preserves finite products and is defined on the category $\mathbb{A}$lgc(k) or on the category $\mathbb{A}$lgcReg(k).

A polynomial $P \in k[X]$ is separable if it has only simple zeros in its splitting field or, what amounts to the same, if it factorises into distinct irreducible polynomials in any algebraic extension of k. An algebraic element of a k-algebra is separable if its minimal polynomial is separable. An algebraic k-algebra is separable if all its elements are separable. Such an algebra is regular since the minimal polynomial P of an element is of the form $P = P_1 \times \ldots \times P_n$ with $P_1,\ldots,P_n$ distinct irreducible polynomials, and as a result, the sub-k-algebra generated by α is $k[\alpha] = k[X]/P \simeq k[X]/P_1 \times \ldots \times k[X]/P_n$, a finite product of fields, thus a regular ring. The category $\mathbb{A}$lgcAlgSep(k) is the full subcategory of $\mathbb{A}$lgc(k) whose objects are the separable algebraic k-algebras. It is also a full subcategory of $\mathbb{A}$lgcReg(k). It is closed under finite products since, if α (resp. β) is a separable element of a k-algebra A (resp. B) with minimal polynomial P (resp. Q), then the polynomial formed as the product of the distinct irreducible factors of P or Q is separable and is the minimal polynomial of the element (α,β) of $A\times B$, and consequently (α,β) is separable in $A\times B$. For each k-algebra A, denote by $S(A)$ the set of separable algebraic elements of A. We will show that $S(A)$ is a sub-k-algebra of A. Let α be a separable element of a k-algebra A. For any algebraic extension field L of k, the minimal polynomial, $P \in k[X]$, of α factorises in $L[X]$ as a product $P = P_1 \times \ldots \times P_n$ of distinct separable irreducible polynomials and consequently the k-algebra, $k[\alpha] \otimes_k L = k[X]/_P \otimes_k L = L[X]/_P \simeq L[X]/P_1 \times \ldots \times L[X]/P_n$ is a separable algebra. Let β be another separable element of A with minimal polynomial $Q = Q_1 \times \ldots \times Q_p$ with $Q_1,\ldots,Q_p$ distinct separable irreducible polynomials. The sub-k-algebra generated by α and β is a quotient of the k-algebra $k[\alpha] \otimes_k k[\beta] = k[\alpha] \otimes_k (k[X]/_Q) \simeq$ $k[\alpha] \otimes_k (k[X]/_{Q_1} \times \ldots \times k|X|/_{Q_p}) \simeq \prod_{j\in[1,p]} k[\alpha] \otimes_k (k[X]/_{Q_j})$, a finite product of separable k-algebras and thus a separable algebra. It follows that $S(A)$ is a separable algebraic sub-k-algebra of A. One thus defines an idempotent endofunctor $S : \mathbb{A}\mathrm{lgc}(k) \to \mathbb{A}\mathrm{lgc}(k)$ and a natural transformation $\varepsilon : S \to 1_{\mathbb{A}\mathrm{lgc}(k)}$ determining an idempotent comonad $\mathbb{S}$ on $\mathbb{A}$lgc(k). The category $\mathbb{A}\mathrm{lgc}^{\mathbb{S}}(k)$ is equivalent to the category $\mathbb{A}$lgcAlgSep(k). One proves, as in 1.12.3., that the comonad $\mathbb{S}$ is of finite rank and that it preserves finite products. It is immediate that $\mathbb{S}$ induces an idempotent comonad of finite rank on $\mathbb{A}$lgcReg(k), whose category of coalgebras is equivalent to $\mathbb{A}$lgcAlgSep(k). It follows that the category $\mathbb{A}$lgcAlgSep(k) is locally simple. ∎

2.13.1.3. <u>The category $\mathbb{A}$lgcBaIntEng(R) of integrally generated commutative Baer R-algebras</u> over an integral domain R, is comonadic for an idempotent comonad of finite rank, preserving finite products, on the category $\mathbb{A}$lgcBa(R) ; it is the full subcategory of $\mathbb{A}$lgcBa(R) having, as its objects, the Baer R-algebras which are generated by their elements which are integral over R. It is a locally simple category. ∎

2.13.1.4. <u>The category $\mathbb{A}$lgcBaStLatIntEng, of strongly lattice-ordered, integrally generated, commutative Baer R-algebras</u> over a totally ordered integral domain R, is comonadic for an idempotent comonad of finite rank, preserving finite products, on the category $\mathbb{A}$lgcBaStLat(R). It is a locally simple category. ■

2.13.1.5. <u>The category $\mathbb{A}$lgcRegStLatAlgEng of strongly lattice-ordered, algebraically generated, commutative k-algebras</u> over a totally ordered commutative field, k, is comonadic for an idempotent comonad of finite rank, preserving finite products, on the category $\mathbb{A}$lgcRegLatReg. It is a locally simple category. ■

2.14. CONTINUOUS REPRESENTATIONS OF PROFINITE GROUPS.

2.14.0. <u>Representations of a group in a category</u>.

2.14.0.0. Given a group G and a category $\mathbb{A}$, <u>a representation of G in $\mathbb{A}$</u> is a pair (A,α) where A is an object of $\mathbb{A}$ and $\alpha : G \to \mathrm{Aut}(A)$ is a group homomorphism from G to the automorphism group of A. A morphism of representations of G in $\mathbb{A}$ with source (A,α) and target (B,β) is a morphism $f : A \to B$ satisfying $f\alpha(\sigma) = \beta(\sigma)f$ for all $\sigma \in G$. Representations of G in $\mathbb{A}$ and their morphisms constitute the category $\mathbb{A}^G$, which is none other than the category of functors from the category G, canonically associated to the group G, to the category $\mathbb{A}$. The forgetful functor $U^G : \mathbb{A}^G \to \mathbb{A}$ is defined by $U^G(A,\alpha) = A$ and $U^G(f) = f$.

If (A,α) is a representation of G in $\mathbb{A}$, the group G operates on the sets $\mathrm{Hom}_{\mathbb{A}}(X,A)$ by $\sigma.u = \alpha(\sigma)u$, which allows one to define the functor $H_{(A,\alpha)} : \mathbb{A}^{op} \to \mathbb{E}ns^G$ by $H_{(A,\alpha)} = \mathrm{Hom}_{\mathbb{A}}(-,A)$ and to define the <u>subgroup of G stabiliser of a morphism</u> $u : X \to A$, by $St(u) = \{\sigma \in G : \alpha(\sigma)u = u\}$.

2.14.0.1. <u>Proposition</u>. <u>If (A,α) is a representation of G in $\mathbb{A}$, the stabiliser subgroups of morphisms with target A satisfy the following properties</u> :

(1) <u>if $u : X \to A$, and $v : Y \to X$, $St(u) \subset St(uv)$</u>.

(2) <u>if $u : X \to A$, and $\mu \in G$, $St(\alpha(\mu)u) = \mu St(u)\mu^{-1}$</u>.

(3) <u>if $u : X \to A$, and</u> $(v_i : Y_i \to X)_{i\in I}$ <u>is an epimorphic</u>

family, $St(u) = \bigcap_{i\in I} St(uv_i)$.

(4) if $u : X \to A$, $v : X \to A$ and $(u,v) : X \to A\times A$, $St(u,v) = St(u) \cap St(v)$.

(5) if $f : (A,\alpha) \to (B,\beta)$ is a morphism of representations of G in $\mathbb{A}$, and $u : X \to A$, then $St(u) \subset St(fu)$; if moreover f is a monomorphism, then $St(u) = St(fu)$.

Proof : (1) $\sigma \in St(u) \Rightarrow \alpha(\sigma)u = u \Rightarrow \alpha(\sigma)uv = uv \Rightarrow \sigma \in St(uv)$.

(2) $\sigma \in St(\alpha(\mu)u) \Leftrightarrow \alpha(\sigma)\alpha(\mu)u = \alpha(\mu)u \Leftrightarrow \alpha(u)^{-1}\alpha(\sigma)\alpha(\mu)u = u$ $\Leftrightarrow \alpha(\mu^{-1}\sigma\mu)u = u \Leftrightarrow \mu^{-1}\sigma\mu \in St(u) \Leftrightarrow \sigma \in \mu\, St(u)\mu^{-1}$.

(3) $\sigma \in St(u) \Leftrightarrow \alpha(\sigma)u = u \Leftrightarrow \forall i \in I,\ \alpha(\sigma)uv_i = uv_i \Leftrightarrow$ $\forall i \in I,\ \sigma \in St(uv_i) \Leftrightarrow \sigma \in \bigcap_{i\in I} St(uv_i)$.

(4) Considering the product representation on $A\times A$, one has : $\sigma \in St(u,v) \Leftrightarrow (\alpha(\sigma)\times\alpha(\sigma))(u,v) = (u,v) \Leftrightarrow \alpha(\sigma)u = u$ and $\alpha(\sigma)v = v \Leftrightarrow \sigma \in St(u)$ and $\sigma \in St(v) \Leftrightarrow \sigma \in St(u) \cap St(v)$.

(5) $\sigma \in St(u) \Leftrightarrow \alpha(\sigma)u = u \Rightarrow f\alpha(\sigma)u = fu \Leftrightarrow \beta(\sigma)fu = fu$ $\Leftrightarrow \sigma \in St(fu)$; if f is a monomorphism, $\sigma \in St(fu) \Rightarrow f\alpha(\sigma)u = fu$ $\Leftrightarrow \alpha(\sigma)u = u \Rightarrow \sigma \in St(u)$. ∎

2.14.0.2. Galois groups.

For any object A of a category $\mathbb{A}$, the pair $(A, 1_{Aut(A)})$ is a representation of the group $Aut(A)$ in $\mathbb{A}$. For each morphism $u : X \to A$, the subgroup of $Aut(A)$ which is the stabiliser of u is the Galois group of u, denoted G_u. The group $Aut(A)$ is identical to the Galois group of the morphism $i_A : Z \to A$ from the initial object Z to A and is called the Galois group of the object A, denoted G_A.

2.14.1. Topological enrichment of locally finitely presentable categories.

We will show that the Galois groups of objects in a locally finitely presentable category canonically have a topological structure. Recall that a topological space is zero-dimensional if it is separated and has a topological base of open closed sets. With continuous maps, they make up the category $\mathbb{T}opz$.

2.14.1.0. Proposition. Every locally finitely presentable category is canonically enriched [33, p.181] over the category $\mathbb{T}opz$ of zero dimensional topological spaces.

Proof : Let $\mathbb{A}$ be a locally finitely presentable category and let A and B be two objects of $\mathbb{A}$. For each pair of morphisms $(u : X \to A,\ v : X \to B)$, let be $H_{u,v} = \{f : A \to B : fu = v\}$. Denoting the initial object of $\mathbb{A}$ by Z, and by $i_C : Z \to C$, the unique morphism, one has $H_{i_A,i_B} = \mathrm{Hom}_{\mathbb{A}}(A,B)$. If $(w : Y \to A,\ t : Y \to B)$ is a second pair of morphisms, the morphism $\langle u,w\rangle : X \coprod Y \to A$ and $\langle v,t\rangle : X \coprod Y \to B$ define a pair of morphisms such that $H_{\langle u,w\rangle,\langle v,t\rangle} = H_{u,v} \cap H_{w,t}$. The family of sets $H_{u,v}$, where (u,v) is a pair of morphisms whose common source is finitely presentable in $\mathbb{A}$, defines a topological base on $\mathrm{Hom}_{\mathbb{A}}(A,B)$. This topology is separated since, if f and g are two distinct morphisms from A to B, there is a finitely presentable object X of $\mathbb{A}$ and a morphism $u : X \to A$ satisfying $fu \neq gu$ and thus $f \in H_{u,fu}$, $g \in H_{u,gu}$ and $H_{u,fu} \cap H_{u,gu} = \emptyset$. The sets $H_{u,v}$ are closed for the topology since if $f : A \to B \notin H_{u,v}$. Then $f \in H_{u,fu}$ with $H_{u,fu} \cap H_{u,v} = \emptyset$ The topology is thus zero dimensional. If C is another object of $\mathbb{A}$, the composition $\Gamma_{A,B,C} : \mathrm{Hom}_{\mathbb{A}}(B,C)\times\mathrm{Hom}_{\mathbb{A}}(A,B) \to \mathrm{Hom}_{\mathbb{A}}(A,C)$ is continuous since, for any pair $(g : B \to C,\ f : A \to B)$ and any basic open set $H_{u,v}$ of $\mathrm{Hom}_{\mathbb{A}}(A,C)$ containing gf, the open set $H_{fu,v}$ of $\mathrm{Hom}_{\mathbb{A}}(B,C)$ contains g, the open set $H_{u,fu}$ of $\mathrm{Hom}_A(A,B)$ contains f and $\Gamma_{A,B,C}(H_{fu,v}\times H_{u,fu}) \subset H_{u,v}$. The bifunctor $\mathrm{Hom}_{\mathbb{A}} : \mathbb{A}^{op}\times\mathbb{A} \to \mathbb{E}ns$ thus lifts to a bifunctor $\mathbb{A}^{op}\times\mathbb{A} \to \mathbb{T}opz$ [33, p.181]. ∎

2.14.1.1. <u>Notation</u>. We will denote by $\mathbb{H}om_{\mathbb{A}}(A,B)$ the set $\mathrm{Hom}_{\mathbb{A}}(A,B)$ with the above defined topology and $\mathbb{H}om_{\mathbb{A}} : \mathbb{A}^{op}\times\mathbb{A} \to \mathbb{T}opz$, the corresponding bifunctor.

2.14.1.2. <u>Proposition</u>. <u>Galois groups of objects of a locally finitely presentable category are zero dimensional topological groups having a base of open neighbourhoods of the identity made up of the Galois groups of morphisms with finitely presentable source</u>.

<u>Proof</u> : For any object A of $\mathbb{A}$, the topological space, $\mathbb{H}om_{\mathbb{A}}(A,A)$ induces on G_A, a zero dimensional topology such that the operations of composition and inversion are continuous, that is, a topology making G_A a topological group. A base of open-closed neighbourhoods of the identity of G_A is made up of subsets of the form, $H_{u,u} \cap G_A = G_u$ (notations 2.14.0.2.). ∎

2.14.1.3. Proposition. The Galois group, G_u, of a morphism $u : X \to A$ is a closed subgroup of the Galois group G_A.

Proof : The object X is a filtered colimit of finitely presentable objects : say $(X,(\iota_i)) = \varinjlim_{i\in\mathbb{I}} X_i$. Then $G_u = \bigcap_{i\in Obj(\mathbb{I})} G u\iota_i$ is a closed subgroup of G_A. ■

2.14.2. Continuous representations of a topological group.

2.14.2.0. Definition. A representation (A,α) of a topological group G in a locally finitely presentable category $\mathbb{A}$, is continuous if the group homomorphism $\alpha : G \to G_A$ is continuous.

The category $\mathbb{A}^{Cont(G)}$ of continuous representations of G in $\mathbb{A}$ is the full subcategory of $\mathbb{A}^G$ whose objects are the continuous representations. The forgetful functor $U^G : \mathbb{A}^{Cont(G)} \to \mathbb{A}$ is the restriction of the functor $U^G : \mathbb{A}^G \to \mathbb{A}$ (cf. 1.14.0.0.).

2.14.2.1. Proposition. If $\mathbb{A}$ is a locally finitely presentable category and G is a topological group, the forgetful functor $U^G : \mathbb{A}^{Cont(G)} \to \mathbb{A}$ from the category of continuous representations of G in $\mathbb{A}$, is comonadic.

Proof : If G is a discrete topological group, the category $\mathbb{A}^{Cont(G)}$ is identical to the category $\mathbb{A}^G$, and the functor $U^G : \mathbb{A}^G \to \mathbb{A}$ has a left adjoint, a right adjoint, and it creates isomorphisms. It is therefore comonadic. In the general case, the functor forgetting structure, $U^G : \mathbb{A}^{Cont(G)} \to \mathbb{A}$ is the composite of the inclusion functor of $\mathbb{A}^{Cont(G)}$ in $\mathbb{A}^G$ and the forgetful functor $U^G : \mathbb{A}^G \to \mathbb{A}$. Let us show that $\mathbb{A}^{Cont(G)}$ is a coreflexive subcategory of $\mathbb{A}^G$. Let (A,α) be an object of $\mathbb{A}^G$. Let us denote M the set of monomorphisms with target A and stabiliser an open subgroup of G. The set M is closed under finite unions, since $St(m \cup n) = St(m) \cap St(n)$ (Proposition 2.14.0.1 (3)) and consequently, it is a filtered ordered set. Let us denote by $\varepsilon_A : LA \to A$, the union of the monomorphisms of M and, for each $m \in M$, let us denote by $\bar{m}$, the monomorphism satisfying $\varepsilon_A \bar{m} = m$. The family of morphisms $(\bar{m} : M \to LA)_{m\in M}$ is extremal epimorphic. Each element $\sigma \in G$ determines, for each $m \in M$, a monomorphism $\alpha(\sigma)m$ with target A whose stabiliser, $St(\alpha(\sigma)m) = \sigma\, St(m)\sigma^{-1}$, (Proposition 2.14.0.1), is open in G, and which, consequently, is factorised by ε_A in a morphism

$\alpha_m(\sigma) : M \to LA$. The family of morphisms $(\alpha_m(\sigma))_{m\in M}$ determines a unique morphism $L(\alpha(\sigma)) : LA \to LA$ satisfying $\varepsilon_A L(\alpha(\sigma))\bar{m} = \alpha(\sigma)m$ for any $m \in M$. We thus obtain the following relation : $\varepsilon_A L(\alpha(\sigma)) = \alpha(\sigma)\varepsilon_A$ which proves that LA is naturally equipped with a representation of G such that ε_A is a morphism of representations. The representation of G on LA is continuous, since any morphism $u : X \to LA$, where X is a finitely presentable object, can be factorised in the form $u = \bar{m}v$ with $m \in M$, and the relations : $St(m) = St(\varepsilon_A \bar{m}) = St(\bar{m}) \subset St(\bar{m}v) = St(u)$, imply that $St(u)$ is open. Let us show that $\varepsilon_A : LA \to A$ is a couniversal morphism from $\mathbb{A}^{Cont(G)}$ to $\mathbb{A}$. Let (B,β) be a continuous representation of G and $f : (B,\beta) \to (A,\alpha)$ a representation morphism. The object B is a filtered colimit of finitely presentable objects of $\mathbb{A}$: say, $(B,(\beta_i)) = \varinjlim_{i\in\mathbb{I}} B_i$. Each morphism $f\beta_i : B_i \to A$ can be factorised in the form $f\beta_i = m_i g_i$, where $g_i : B_i \to M_i$ is an extremal epimorphism and m_i is a monomorphism [16]. The stabiliser of β_i is open and the relations : $St(m_i) = St(m_i g_i) = St(f\beta_i) \supset St(\beta_i)$, imply that the stabiliser of m_i is open. It follows that m_i is factorised by ε_A, therefore that the morphisms $f\beta_i$ are factorised by ε_A, and consequently, that f is factorised by ε_A giving a representation morphism of G. This proves that ε_A is couniversal and that $\mathbb{A}^{Cont(G)}$ is a coreflexive subcategory of $\mathbb{A}^G$. It is immediate that $\mathbb{A}^{Cont(G)}$ is closed in $\mathbb{A}^G$ for subobjects and consequently that $\mathbb{A}^{Cont(G)}$ has equalisers which are preserved by the inclusion functor of $\mathbb{A}^{Cont(G)}$ into $\mathbb{A}^G$. The functor forgetting the action, $U^G : \mathbb{A}^{Cont(G)} \to \mathbb{A}$, therefore, preserves equalisers and, since it creates isomorphisms, it is comonadic by Beck's theorem [33, p.147]. ■

2.14.3. Continuous representations of a profinite group. Let us remember that a profinite group is a topological group which is a projective limit of finite discrete groups or, equivalently, a separated compact topological group with a base of neighbourhoods of the identity element formed by a set of normal open subgroups stable under finite intersections, or, also , a topological group whose underlying topological space is Boolean, [23, theorem 8.4.1.].

2.14.3.0. Theorem. If G is a profinite group and $\mathbb{A}$ is a locally finitely presentable (resp. locally indecomposable, resp. locally simple) category, the category $\mathbb{A}^{Cont(G)}$ of continuous representations of G in $\mathbb{A}$ is a locally finitely presentable (resp. locally indecomposable, resp. locally simple) category, which is comonadic over the category $\mathbb{A}$ for a comonad of finite rank.

<u>Proof</u> : According to Proposition 2.14.2.1., the forgetful functor $U^G : \mathbb{A}^{Cont(G)} \to \mathbb{A}$ is comonadic and generates a comonad on $\mathbb{A}$, which we will denote by $\mathbb{L} = (L,\varepsilon,\delta)$. Let us show that $\mathbb{L}$ is of finite rank. Let X be a finitely presentable object of $\mathbb{A}$, (A,α) a continuous representation of G in $\mathbb{A}$ and $f : X \to A$ a morphism of $\mathbb{A}$. The subgroup $St(f)$ is open in G, therefore the homogeneous G-space $G/St(f)$ of left cosets of G modulo $St(f)$ is finite. Let $\{\sigma_1,\dots,\sigma_n\}$ be a set of representatives for these cosets ; it is an homogeneous G-space. For any element μ of G, we define a permutation $\bar{\mu}$ of $\{1,\dots,n\}$ by writing $\sigma_{\bar{\mu}(i)} = \mu.\sigma_i$. For an element ν of G, we have $\sigma_{\overline{\nu\mu}(i)} = (\nu\mu).\sigma_i = \nu.(\mu.\sigma_i) = \nu.\sigma_{\bar{\mu}(i)} = \sigma_{\bar{\nu}(\bar{\mu}(i))}$ and consequently $\overline{\nu\mu} = \bar{\nu} \circ \bar{\mu}$. We denote by $(B, (\iota_i)) = \coprod_n X$, the sum of n copies of the object X. It is a finitely presentable object of $\mathbb{A}$. For each $\mu \in G$, let us define the morphism $\beta(\mu)$ by $\beta(\mu) = \langle\iota_{\bar{\mu}(i)}\rangle_{i\in[1,n]} : \coprod_n X \to \coprod_n X$. For $\nu \in G$, we have : $\beta(\nu\mu) = \langle\iota_{\overline{\nu\mu}(i)}\rangle = \langle\iota_{\bar{\nu}(\bar{\mu}(i))}\rangle = \beta(\nu)\,\langle\iota_{\bar{\mu}(i)}\rangle = \beta(\nu)\,\beta(\mu)$. It follows that (B,β) is a representation of G in $\mathbb{A}$. For any $\sigma \in St(f)$, we have : $\sigma.\sigma_i = \sigma_i$ for any $i\in[1,n]$, therefore $\bar{\sigma} = \bar{1}$ and $\beta(\sigma)=1_B$. The representation (B,β) is continuous as the stabiliser subgroup of any morphism with target B contains $St(f)$ and therefore is open. The morphism $\langle\alpha(\sigma_i)f\rangle_{i\in[1,n]} : \coprod_n X \to A$ is a morphism of representations: $(B,\beta) \to (A,\alpha)$ since, for any $\mu \in G$ and any $i \in [1,n]$, the relation $\mu.\sigma_i = \sigma_{\bar{\mu}(i)}$ implies $\mu\sigma_i \sim \sigma_{\bar{\mu}(i)}$ (modulo $St(f)$), therefore $\alpha(\mu\sigma_i)f = \alpha(\sigma_{\bar{\mu}(i)})f$ and consequently $\alpha(\mu)\,\langle\alpha(\sigma_i)f\rangle\iota_i = \alpha(\mu)\,\alpha(\sigma_i)f = \alpha(\mu\sigma_i)f = \alpha(\sigma_{\bar{\mu}(i)})f = \langle\alpha(\sigma_i)f\rangle\iota_{\bar{\mu}(i)} = \langle\alpha(\sigma_i)f\rangle\,\beta(\mu)\iota_i$, that is to say $\alpha(\mu)\,\langle\alpha(\sigma_i)f\rangle = \langle\alpha(\sigma_i)f\rangle\,\beta(u)$. As a result, any morphism $f : X \to A$ from a finitely presentable object X to an object A equipped with a continuous representation of G, can be factorised in the form $f = hg$ where $g : X \to B$ is a morphism whose target B is finitely presentable and $h : (B,\beta) \to (A,\alpha)$ is a morphism of continuous representations of G.

Let us prove that the functor $L : \mathbb{A} \to \mathbb{A}$ preserves filtered colimits. Let $(\iota_i : A_i \to A)_{i\in\mathbb{I}}$ be a filtered colimit in $\mathbb{A}$, $(\psi_i : LA_i \to D)_{i\in\mathbb{I}}$ the filtered colimit of the diagram $(LA_i)_{i\in\mathbb{I}}$ and $d : D \to LA$ the morphism satisfying the relation $d\psi_i = L\iota_i$ for each $i \in \mathbb{I}$. The morphism $\varepsilon_A d : D \to A$ is a filtered colimit of the monomorphisms ε_{A_i} $(i \in \mathbb{I})$. It is a monomorphism, [16], therefore d is a monomorphism. Let X be a finitely presentable object of $\mathbb{A}$ and $f : X \to LA$ a morphism. Since LA is equipped with a continuous representation of G, there exists a morphism of continuous representations, $h : (B,\beta) \to (LA,\delta_A)$, with B finitely presentable, and a morphism $g : X \to B$ satisfying $f = hg$. The morphism $\varepsilon_A h$ factorises in the form $\varepsilon_A h = \iota_i f_i = \iota_i \varepsilon_{A_i} g_i$. The relations : $\varepsilon_A f = \varepsilon_A hg = \iota_i f_i g = \iota_i \varepsilon_{A_i} g_i g = \varepsilon_A (L\iota_i) g_i g = \varepsilon_A d\psi_i g_i g$ imply the relation : $f = d\psi_i g_i g$, and thus the factorisation of f through d. It follows that d is an isomorphism and consequently that L preserves filtered colimits. One can conclude that the comonad $\mathbb{L}$, generated by the functor $U^G : \mathbb{A}^{Cont(G)} \to \mathbb{A}$, is of finite rank (definition 1.12.0). It follows that the category $\mathbb{A}^{Cont(G)}$ is locally finitely presentable (Proposition 1.12.1.). The category $\mathbb{A}^{Cont(G)}$ is closed in $\mathbb{A}^G$ under finite products, therefore the functor $U^G : \mathbb{A}^{Cont(G)} \to \mathbb{A}$ preserves finite products as does the functor L. It follows from Proposition 1.12.2. (resp. 2.13.0.) that the category $\mathbb{A}^{Cont(G)}$ is locally indecomposable (resp. locally simple) if the category $\mathbb{A}$ is. ■

2.14.3.1. Examples. If G is a profinite group, the category $\mathbb{B}ool^{Cont(G)}$ of continuous representations of G in the category of Boolean algebras, is a locally simple category equivalent to the dual of the category, G-$\mathbb{E}$spBool, of Boolean topological G-spaces.

Proof : The category $\mathbb{B}$ool being locally simple, the category $\mathbb{B}ool^{Cont(G)}$ is locally simple and its finitely presentable objects are the continuous representations of G on finite Boolean algebras. (Proposition 1.12.1.). Let $(X,.)$ be a Boolean G-space. We denote by $B(X)$, the Boolean algebra of open-closed subsets of X. The pair $(B(X),\beta_X)$ defined by $\beta_X(u) = \sigma.u = \{\sigma.x : x \in u\}$ is a representation of G in $B(X)$. Let us show that for $u \in B(X)$, $St(u) = \{\sigma\in G : \sigma.u = u\}$ is open in G. The multiplication $G\times X \to X$ being continuous, for each element x of u, there is an open subgroup H_x and an open subset u_x of X containing x such that $H_x.u_x \subset u$. From the covering of u by the open sets u_x, one can extract a finite covering $(u_{x_i})_{i\in[1,n]}$. The open subgroup $H = \bigcap_{i\in[1,n]} H_{x_i}$ of G is such that

$H.u = (\bigcap_{i\in[1,n]} H_{x_i}).(\bigcup_{j\in[1,n]} u_{x_j}) \subset \bigcup_{i\in[1,n]} (H_{x_i}.u_{x_i}) \subset u$. Then $St(u) \supset H$ is an open subgroup of G. Let B_o be a finitely presentable object of $\mathbb{B}ool$ and $f : B_o \to B(X)$ a morphism of $\mathbb{B}ool$. The Boolean algebra B_o being finite, $St(f) = \bigcap_{u\in f(B_o)} St(u)$ is an open subgroup of G. Consequently $(B(X),\beta_X)$ is an object of $\mathbb{B}ool^{Cont(G)}$. For any morphism $f : (X,.) \to (Y,.)$ of G-$\mathbb{E}sp\mathbb{B}ool$, the morphism $B(f) : (B(Y),\beta_Y) \to (B(X),\beta_X)$ is defined by $B(f)(v) = f^{-1}(v)$. In this way, we obtain a functor $B : G - \mathbb{E}sp\mathbb{B}ool^{op} \to \mathbb{B}ool^{Cont(G)}$. This functor is fully faithful since continuous maps : $X \to Y$, correspond bijectively to Boolean algebraic homomorphisms : $B(Y) \to B(X)$ and, consequently the morphisms $(X,.) \to (Y,.)$ correspond bijectively to morphisms $(B(Y),\beta_Y) \to (B(X),\beta_X)$. Let us show that this functor is essentially surjective. For any object (B,α) of $\mathbb{B}ool^{Cont(G)}$, there exists a Boolean space X such as $B(X) \simeq B$, and for any $\sigma \in G$, the isomorphism of Boolean algebras $\alpha(\sigma) : B \to B$ determines a unique homeomorphism $X \to X$, whose inverse homeomorphism will be denoted $\mu(\sigma)$. Then (X,μ) is a G-set. It is a G-space since, for any $(\sigma,x) \in G\times X$ and any open-closed u of X such that $\sigma.x \in u$, the open set $\sigma.St(\sigma^{-1}u) \times \sigma^{-1}(u)$ of $G\times X$ contains (σ,x) and is such that $\sigma.St(\sigma^{-1}u).\sigma^{-1}(u) = \sigma.(St(\sigma^{-1}u).\sigma^{-1}u) = \sigma.\sigma^{-1}u = u$. It is immediate that $B(X,\mu) \simeq (B,\alpha)$. ∎

2.15. LOCALLY NOETHERIAN LOCALLY SIMPLE CATEGORIES.

Following to P. Gabriel and F. Ulmer [16], an object A of a category $\mathbb{A}$ will be said to be Noetherian if any decreasing sequence of extremal quotient objects of A is stationary, and a category is locally Noetherian if it is cocomplete and has a generating set formed of finitely presentable Noetherian objects, finite sums of which are still Noetherian objects, or, in an equivalent way, if it is a locally finitely presentable category in which any finitely generated objects is finitely presentable. An object is said to be semi-simple if it is a finite product of simple objects.

2.15.0. Proposition. In a locally indecomposable category, semi-simple objects are Noetherian and are stable for extremal quotients.

Proof : Let $(S,(p_i)) = \prod_{i\in[1,n]} S_i$ be a finite product of simple objects and $q : S \to T$ an extremal epimorphism. For each $i \in [1,n]$, we denote by $(t_i : T \to T_i,\ q_i : S_i \to T_i)$, the amalgamated sum of (q,p_i). Since finite products are couniversal, the family $(t_i : T \to T_i)_{i\in[1,n]}$ is a finite product and $q = \prod_{i\in[1,n]} q_i$. Each morphism q_i, being an extremal epimorphism, either is an isomorphism or has target 1. Writing $I = \{i \in [1,n] : q_i$ is an isomorphism$\}$, the morphism q is isomorphic to the canonical projection $\prod_{i\in[1,n]} S_i \to \prod_{i\in I} S_i$. The object $T \simeq \prod_{i\in I} S_i$ is therefore semi-simple. Moreover, the object S is Noetherian as it only has a finite set of extremal quotient objects. ∎

2.15.1. Proposition. A category is locally Noetherian locally simple if, and only if, it is a locally indecomposable category the finitely presentable objects of which are semi-simple.

Proof : Let A be a finitely presentable object of a locally Noetherian locally simple category $\mathbb{A}$. The object A is a Noetherian object [16]. The Boolean algebra $\Delta(A)$ of direct factors of A is Noetherian, therefore finite. The maximal spectrum $\mathrm{Spec}_{\mathrm{Max}}(A)$ is therefore a finite topological space, and thus it is discrete. The object A is the object of global sections of the structural sheaf, $\tilde{A}$, over $\mathrm{Spec}_{\mathrm{Max}}(A)$, therefore is a finite product of its simple components, i.e., it is a semi-simple object. Conversely, if $\mathbb{A}$ is a locally indecomposable category, the finitely presentable objects of which are semi-simple, then these are Noetherian (Proposition 2.15.0) and codecidable, since such an object has a codiagonal which is a regular quotient object of a semi-simple object, and therefore is a canonical projection of a product (cf. Proof of Proposition 2.15.0) and, hence is complemented. Thus the category $\mathbb{A}$ is locally simple (Proposition 2.1.2). ∎

2.15.2. Proposition. If $\mathbb{A}$ is a locally Noetherian category and $\mathbb{L}$ is a comonad of finite rank over $\mathbb{A}$, the category $\mathbb{A}^{\mathbb{L}}$ of $\mathbb{L}$-coalgebras is locally Noetherian.

Proof : The category $\mathbb{A}^{\mathbb{L}}$ is locally finitely presentable (Proposition 1.12.1) and the forgetful functor $U^{\mathbb{L}} : \mathbb{A}^{\mathbb{L}} \to \mathbb{A}$ preserves colimits thus coequalisers and regular quotient objects, as well as the extremal quotient objects which are filtered colimits of regular quotient objects, [16]. If (A,α) is a finitely presentable object of $\mathbb{A}^{\mathbb{L}}$, then A is a finitely presentable object of $\mathbb{A}$ (Proposition 1.2.1), and therefore is a Noetherian object of $\mathbb{A}$. Any decreasing sequence of extremal quotient objects of (A,α) has, as its image by $U^{\mathbb{L}}$, a decreasing sequence of extremal quotient objects of A. Since this latter sequence is stationary and the functor $U^{\mathbb{L}}$ reflects isomorphisms, the first sequence is stationary. The object (A,a) is therefore Noetherian. This implies that category $\mathbb{A}^{\mathbb{L}}$ is locally Noetherian. ■

2.15.3. Proposition. The category of continuous representations of a profinite group in a locally Noetherian category is locally Noetherian.

Proof : This is a consequence of Theorem 2.14.3.0 and of Proposition 2.15.2. ■

2.15.4. Examples.

2.15.4.0. $\mathbb{A}$lgc(k). If k is a commutative field, the category $\mathbb{A}$lgc(k) is locally Noetherian [16].

2.15.4.1. $\mathbb{A}$lgcAlgReg(k). The category $\mathbb{A}$lgcAlgReg(k) is locally simple (Proposition 2.13.1.1). Let us consider, first, regular algebraic k-algebras generated by a single element. They are of the form $k[\alpha]$ where α is an element whose minimal polynomial P can be factorised as a product of distinct, irreducible polynomials $P_1,\dots,P_n$. Then $k[\alpha] \simeq k[X]/P \simeq k[X]/P_1 \times \dots \times k[X]/P_n$ is a semi-simple k-algebra, and therefore is Noetherian (Proposition 2.15.0). For any algebraic extension field K of k, the polynomial $P \in K[X]$ factorises in the form $P = Q_1^{k_1} \times \dots \times Q_p^{k_p}$ and the semi-simple algebra $K[X]/Q_1 \times \dots \times K[X]/Q_p$ can be seen to be the sum, in the category $\mathbb{A}$lgcReg(k), therefore in $\mathbb{A}$lgcAlgReg(k), of the two k-algebras, $k[\alpha]$ and K, if we verify, first, the universal property for extensions fields of K, and then, use the fact that the class of extension fields of k properly cogenerates the category $\mathbb{A}$lgcAlgReg(k). It then follows from the distributivity of finite sums over finite products, that the sum of the two k-algebras $k[\alpha]$ and $k[\beta]$ in $\mathbb{A}$lgcAlgReg(k) is a semi-simple k-algebra and therefore is Noetherian. The category $\mathbb{A}$lgcAlgReg(k) is therefore locally Noetherian since the k-algebras $k[\alpha]$ properly generate it.

2.15.4.2. $\mathbb{A}$lgcAlgSep(k). The category, $\mathbb{A}$lgcAlgSep(k), is locally Noetherian, since it is equivalent to the category of coalgebras for a comonad of finite rank which preserves finite products, on the locally Noetherian category $\mathbb{A}lgc(k)$ (2.13.1.2).

2.15.5. *Locally Noetherian locally simple categories of functors.*

2.15.5.0. *Notations*. Following [10, definition 1.2.0] , a *multisum* of a family $(X_i)_{i\in I}$ of objects of a category $\mathbb{K}$ is a family $(\gamma_{ji} : X_i \to Y_j)_{(i,j)\in I\times J}$ of morphisms of $\mathbb{K}$ such that, for any family $(f_i : X_i \to Y)_{i\in I}$ of morphisms of $\mathbb{K}$, there exists a unique pair (j,f) made up of an element $j \in J$ and a morphism $f : Y_j \to Y$ satisfying, for all $i \in I$, $f\gamma_{ji} = f_i$. A functor $F : \mathbb{K}^{op} \to \mathbb{E}ns$ is *multicontinuous* for this multisum, if the canonical mapping $\langle(F\gamma_{ji})\rangle : \coprod_{j\in J} FY_j \to \prod_{i\in I} FX_i$ is bijective. The multisum $(\gamma_{ji} : X_i \to Y_j)_{(i,j)\in I\times J}$ is said to be *well finite* if the sets I and J are finite. The category $\mathbb{K}$ has *well finite multisums* if any finite family of objects of $\mathbb{K}$ has a well finite multisum. The functor $F : \mathbb{K}^{op} \to \mathbb{E}ns$ is *finitely multicontinuous*, if it is multicontinuous for all the well finite multisums of $\mathbb{K}$. The *category* $\mathbb{F}amfin\,\mathbb{K}$ of finite families of objects of $\mathbb{K}$ has, as objects, the finite families $(X_i)_{i\in I}$ of objects of $\mathbb{K}$ and, as morphisms from $(X_i)_{i\in I}$ to $(Y_j)_{j\in J}$, the pairs $(\alpha,(f_j)_{j\in J})$ consisting of a mapping $\alpha : J \to I$ and a family of morphisms $f_j : X_{\alpha(j)} \to Y_j$ $(j \in J)$, the composition of morphisms being defined by $(\beta,(g_k)_{k\in K})(\alpha,(f_j)_{j\in J}) = (\alpha\beta,(g_k f_{\beta(k)})_{k\in K})$. The *category* $\mathbb{K}$ is said to be *monomorphic* if all its morphisms are monomorphisms.

2.15.5.1. *Theorem*. *If $\mathbb{K}$ is a small monomorphic category with well finite multisums, the category $\mathbb{C}ont_{\aleph_0}[\mathbb{F}amfin\,\mathbb{K}^{op},\mathbb{E}ns]$ of finitely continuous contravariant functors defined on the category of finite families of objects of $\mathbb{K}$ with values in the category of sets, is locally Noetherian locally simple*.

Proof : The category $\mathbb{K}$ has multiequalisers [10, 1.4.0.], since the multiequaliser of a pair of distinct morphisms $(f,g) : X \rightrightarrows Y$ is empty and the multiequaliser of $(f,f) : X \rightrightarrows Y$ is 1_Y. The category $\mathbb{K}$ is therefore finitely multicocomplete [10, Proposition 1.5.3.]. Let us show that the category $\mathbb{F}amfin\,\mathbb{K}$ is finitely cocomplete. Let

$((X_i)_{i\in I_m})_{m\in\mathbb{M}}$ be a finite diagram of $\mathbb{F}\mathrm{amfin}\,\mathbb{K}$. Write $I = \varprojlim_{m\in\mathbb{M}} I_m$. For any element $i = (i_m)_{m\in\mathbb{M}}$ of I, we denote by $(\iota_{ji_m} : X_{i_m} \to Y_j)_{(m,j)\in\mathbb{M}\times J_i}$ a multicolimit of $(X_{i_m})_{m\in\mathbb{M}}$. The sets J_i are finite. We will suppose them to be pairwise disjoint and we write $J = \bigcup_{i\in I} J_i$. For $m \in \mathbb{M}$, we define $\alpha_m : J \to I_m$ by $\alpha_m(j) = i_m$ if $j \in J_i$. We then obtain a morphism $(\alpha_m)(\iota_{ji_m})_{j\in J}) : (X_i)_{i\in I_m} \to (Y_j)_{j\in J}$ and an inductive cone in $\mathbb{F}\mathrm{amfin}\,\mathbb{K}$, with base $((X_i)_{i\in I_m})_{m\in\mathbb{M}}$ and vertex $(Y_j)_{j\in J}$ that can easily be shown to be a colimit. The category $\mathbb{F}\mathrm{amfin}\,\mathbb{K}$ has also finite products, since any finite family $((X_i)_{i\in I_m})_{m\in M}$ of objects of $\mathbb{F}\mathrm{amfin}\,\mathbb{K}$ in which the sets I_m are supposed pairwise disjoint, has as a product, the object $(X_i)_{i\in I}$ with $I = \bigcup_{m\in M} I_m$, with projections $(X_i)_{i\in I} \to (X_i)_{i\in I_m}$. The products are codisjoint. They are also couniversal since the direct image by a morphism $(\alpha,(f_k)) : (X_i)_{i\in I} \to (Y_k)_{k\in K}$ of the projection $(X_i)_{i\in I} \to (X_i)_{i\in I_m}$ is the projection $(Y_k)_{k\in K} \to (Y_k)_{k\in\alpha^{-1}(I_m)}$ and the family of these images form a product. Let $(X_i)_{i\in I}$ be an object of $\mathbb{F}\mathrm{amfin}\,\mathbb{K}$. For each $i \in I$, let $(\gamma_{j1} : X_i \to Y_j,\ \gamma_{j2} : X_i \to Y_j)_{j\in J_i}$ be the multisum of X_i with itself. There exists a unique $j_i \in J_i$ and a unique morphism, $f_i : Y_{j_i} \to X_i$ satisfying $f_i\gamma_{j_i1} = 1_{X_i} = f_i\gamma_{j_i2}$. The morphism f_i, being a monomorphism, is in fact an isomorphism, the same being true for the morphisms γ_{j_i1} and γ_{j_i2}. One can thus assume that $\gamma_{j_i1} = 1_{X_i} = \gamma_{j_i2}$, $i\epsilon J_i$ and $\gamma_{i1} = 1_{X_i} = \gamma_{i2}$. Taking the sets J_i to be pairwise disjoint, we set $J = \bigcup_{i\in I} j_i$. Then $(\gamma_j)_{j\in J}$ is the sum of the object $(X_i)_{i\in I}$ with itself and the associated codiagonal is the projection $(Y_j)_{j\in J} \to (Y_i)_{i\in I}$, which is therefore a complemented quotient object. The objects of $\mathbb{F}\mathrm{amfin}\,\mathbb{K}$ are therefore codecidable. As a result, $\mathbb{F}\mathrm{amfin}\,\mathbb{K}^{op}$ is a small finitely complete category with finite, universal, disjoint sums and decidable objects. According to Theorem 2.3.0, the category $\mathbb{A} = \mathbb{C}\mathrm{ont}_{\aleph_0}[\mathbb{F}\mathrm{amfin}\,\mathbb{K}^{op},\mathbb{E}\mathrm{ns}]$ of finitely continuous functors defined on $\mathbb{F}\mathrm{amfin}\,\mathbb{K}^{op}$ with values in the category of sets, is locally simple. The category $\mathbb{F}\mathrm{amfin}\,\mathbb{K}$ can be identified with the category of finitely presentable objects of $\mathbb{A}$, and $\mathbb{K}$ with the full subcategory of $\mathbb{F}\mathrm{amfin}\,\mathbb{K}$ having, as objects, the families of objects with only one member. Since the morphisms of $\mathbb{K}$ are monomorphsms, the objects of $\mathbb{K}$ are simple in the categories $\mathbb{K}$, $\mathbb{F}$amfin $\mathbb{K}$ and $\mathbb{A}$. They are finitely presentable simple objects in $\mathbb{A}$. Conversely, any finitely presentable simple object of $\mathbb{A}$ is necessarily an object of $\mathbb{K}$. The category $\mathbb{K}$ is therefore the category of finitely presentable simple objects of $\mathbb{A}$, and the category $\mathbb{F}$amfin $\mathbb{K}$ is the category of semi-simple objects of $\mathbb{A}$. It follows that the category $\mathbb{A}$ is locally Noetherian locally simple (Proposition 2.15.1.). ∎

2.15.5.2. Proposition. The category of simple objects of $\mathbb{C}ont_{\aleph_0}$ [$\mathbb{F}$amfin $\mathbb{K}^{op}$, $\mathbb{E}$ns] is equivalent to the category $\mathbb{M}$ulcont$_{\aleph_0}$ [$\mathbb{K}^{op}$, $\mathbb{E}$ns] of finitely multicontinuous contravariant functors defined on $\mathbb{K}$, and the category of finitely presentable simple objects of $\mathbb{C}ont_{\aleph_0}$ [$\mathbb{F}$amfin $\mathbb{K}^{op}$, $\mathbb{E}$ns] is equivalent to $\mathbb{K}$.

Proof : That the category $\mathbb{K}$ is equivalent to the category of finitely presentable simple objects of $\mathbb{A} = \mathbb{C}ont_{\aleph_0}$ [$\mathbb{F}$amfin $\mathbb{K}^{op}$, $\mathbb{E}$ns] is shown in proof 2.15.5.1. According to Proposition 1.7.0., the category $\mathbb{S}$im $\mathbb{A}$ = $\mathbb{I}$nd $\mathbb{A}$ is locally $\aleph_0$-multipresentable and the category of its finitely presentable objects is equivalent to $\mathbb{K}$. The equivalence $\mathbb{S}$im $\mathbb{A} \sim \mathbb{M}$ulcont$_{\aleph_0}$ [$\mathbb{K}^{op}$, $\mathbb{E}$ns] results from Theorem 4.8. [11]. ∎

2.15.5.3. Examples.

2.15.5.3.0. $\mathbb{K} = \mathbb{1}$.

The category $\mathbb{F}$amfin $\mathbb{1}$ is equivalent to the category $\mathbb{E}$nsfin^{op}, dual to the category of finite sets, and the category $\mathbb{C}ont_{\aleph_0}$ [$\mathbb{F}$amfin $\mathbb{1}^{op}$, $\mathbb{E}$ns] is equivalent to the category, $\mathbb{B}$ool, of Boolean algebras.

2.15.5.3.1. $\mathbb{K}$ = $\mathbb{E}$nsfinMono .

Category of finite sets and injective mappings. The category $\mathbb{M}$ultcont$_{\aleph_0}$ [$\mathbb{K}^{op}$, $\mathbb{E}$ns] is equivalent to the category $\mathbb{E}$nsMono of sets and injective mappings. The category $\mathbb{C}ont_{\aleph_0}$ [$\mathbb{F}$amfin $\mathbb{K}^{op}$, $\mathbb{E}$ns] is then a category of Boolean sheaves of sets.

2.15.5.3.2. $\mathbb{K}$ = $\mathbb{A}$lgfinMono($\mathbb{T}$) .

Let $\mathbb{T}$ be an algebraic theory in the sense of Lawvere, which is locally finite i.e. such that the sets $\mathrm{Hom}_{\mathbb{T}}(n,1)$ are finite. The category $\mathbb{A}$lgfinMono($\mathbb{T}$) has, as objects the finite $\mathbb{T}$-algebras and as morphisms the monomorphisms of $\mathbb{T}$-algebras. The category $\mathbb{M}ultcont_{\aleph_0}[\mathbb{K}^{op}, \mathbb{E}ns]$ is equivalent to the category $\mathbb{A}lgMono(\mathbb{T})$ of $\mathbb{T}$-algebras and $\mathbb{T}$-monomorphisms. The category $\mathbb{C}ont_{\aleph_0'}[\mathbb{F}amfin\,\mathbb{K}^{op}, \mathbb{E}ns]$ is then a category of Boolean sheaves of $\mathbb{T}$-algebras.

2.15.5.3.3. $\underline{\mathbb{K} = \mathbb{O}rdStfin}$.
Category of finite ordered sets and strictly increasing mappings.

2.15.5.3.4. $\underline{\mathbb{K} = \mathbb{O}rdTotfin}$.
Category of totally ordered, finite sets and strictly increasing mappings.

2.15.5.3.5. $\underline{\mathbb{K} = \mathbb{E}xtAlgfin(k)}$.
Category of algebraic extension fields of finite degree of a commutative field k. It is a small monomorphic category. Let us show that it has well finite multisums. Let L and M be two algebraic extensions of finite degree of k. We are going to show by induction on the degree of L that the well finite multisum of L and M exists. It is immediate if degree (L) = 1. Let L be of degree n. There exists an intermediate field K between k and L such that K is an algebraic extension of k of degree less than n, and L is an algebraic extension of K of the form $L = K[\alpha]$ where α is algebraic over K with minimal polynomial $P(X) \in K[X]$. The well finite multisum of K and M exists and consists of a finite family $(N_i)_{i\in I}$ of algebraic extensions of finite degree of K and M. For each $i \in I$, the polynomial $P(X) \in N_i[X]$ factorises in the form $P(X) = \prod_{j\in J_i} P_j^{k_j}(X)$ in which the polynomials $P_j(X)$ are irreducible and distinct. We check that the finite family $(N_i[X]/P_j(X))_{(i,j) \in \coprod_{i\in I} J_i}$ of algebraic extensions of L and M gives a well finite multisum for L and M in $\mathbb{E}xtAlgfin(k)$. The category $\mathbb{M}ultcont_{\aleph_0}[\mathbb{K}^{op}, \mathbb{E}ns]$ is equivalent to the category $\mathbb{E}xtAlg(k)$ of algebraic extension fields of k, and the category $\mathbb{C}ont_{\aleph_0'}[\mathbb{F}amfin\,\mathbb{K}^{op}, \mathbb{E}ns]$ is equivalent to the category $\mathbb{A}lgcAlgReg(k)$, of regular algebraic k-algebras.

2.15.5.3.6. $\underline{\mathbb{K} = \mathbb{E}xtAlgfinSep(k)}$: category of separable algebraic extension fields of finite degree of a commutative field k. The category $\mathbb{M}ulcont_{\aleph_0'}[\mathbb{K}^{op}, \mathbb{E}ns]$ is equivalent to the category

$\mathbb{E}$xtAlgSep(k) of separable algebraic extension fields of k, and the category $\mathbb{C}\mathrm{ont}_{\aleph_0}$ [$\mathbb{F}$amfin $\mathbb{K}^{op}$, $\mathbb{E}$ns] is equivalent to the category k-$\mathbb{A}$lgcAlgSep(k) of separable algebraic commutative k-algebras.

2.15.6. Theorem. Any locally Noetherian locally simple category is equivalent to the category $\mathbb{C}\mathrm{ont}_{\aleph_0}$ [$\mathbb{F}$amfin $\mathbb{K}^{op}$, $\mathbb{E}$ns] of finitely continuous contravariant functors on the category of finite families of objects of a small monomorphic category with well finite multisums, $\mathbb{K}$.

Proof : Let $\mathbb{A}$ be a locally Noetherian locally simple category. We denote by $\mathbb{A}_0$ the full subcategory of $\mathbb{A}$ having, as objects, the finitely presentable objects of $\mathbb{A}$, and we denote by $\mathbb{K}$, the full subcategory of $\mathbb{A}_0$ having as objects, the finitely presentable simple objects of $\mathbb{A}$. $\mathbb{K}$ is a small monomorphic category. Let $(X_i)_{i\in[1,n]}$ be a finite family of objects of $\mathbb{K}$ with sum $(\coprod_{i\in[1,n]}X_i,(\iota_i))$ in $\mathbb{A}$. The object $\coprod_{i\in[1,n]}X_i$ is finitely presentable in $\mathbb{A}$, therefore semi-simple (Proposition 2.15.1) and there is a finite family $(Y_j)_{j\in[1,p]}$ of objects of $\mathbb{K}$ with product $(\coprod_{i\in[i,n]}X_i,(p_j))$. It is immediate that the family $(p_j\iota_i : X_i \to Y_j)_{(i,j)\in[1,n]\times[i,p]}$ is a multisum of $(X_i)_{i\in[1,n]}$ in $\mathbb{K}$. The category $\mathbb{K}$ therefore has well finite multisums. Consider the functor $P : \mathbb{F}\mathrm{amfin}\,\mathbb{K} \to \mathbb{A}_0$ defined, on an object $(X_i)_{i\in[1,n]}$, by $P((X_i)_{i\in[1,n]}) = \prod_{i\in[1,n]}X_i$ and, on a morphism $(\alpha,(f_j)) : (X_i)_{i\in[1,n]} \to (Y_j)_{j\in[1,p]}$, by $P(\alpha,(f_j)) = (f_jp_{\alpha(j)})_{j\in[1,p]}$ where $p_i : \prod_{i\in[1,n]}X_i \to X_i$ is the canonical projection of index i. The functor P is faithful since, if two morphisms $(\alpha,(f_j))$ and $(\beta,(g_j))$ from $(X_i)_{i\in[1,n]}$ to $(Y_j)_{j\in[1,p]}$ satisfy $P(\alpha,(f_j)) = P(\beta,(g_j))$, the relations $f_jp_{\alpha(j)} = g_jp_{\beta(j)}$ together with the fact that the product $\prod_{i\in[1,n]}X_i$ is codisjoint and that the objects Y_j are simple, imply that $\alpha(j) = \beta(j)$ and $f_j = g_j$. The functor P is full since, for any morphism $f : P(X_i)_{i\in[1,n]} \to P(Y_j)_{j\in[1,p]}$ and any $j \in [i,p]$, the morphism $p_jf : \prod_{i\in[1,n]}X_i \to Y_j$ with Y_j simple, factorises in a unique way in the form $p_jf = f_jp_{\alpha(j)}$, which determines a function $\alpha : [1,p] \to [1,n]$ and a family of morphisms $(f_j)_{j\in[1,p]}$ defining a morphism $(\alpha,(f_j)) : (X_i)_{i\in[1,n]} \to (Y_j)_{j\in[1,p]}$ such that $P(\alpha,(f_j)) = f$. The functor P is essentially surjective, since any object of $\mathbb{A}_0$ is a finite product of finitely presentable simple objects of $\mathbb{A}$ (Proposition 2.15.1.). The functor P is therefore an equivalence of categories. As a result, we have the following equivalences

$\mathbb{A} \sim \mathbb{C}ont_{\aleph_0}[\mathbb{A}_0^{op}, \mathbb{E}ns] \sim \mathbb{C}ont_{\aleph_0}[\mathbb{F}amfin\,\mathbb{K}^{op}, \mathbb{E}ns]$. ∎

2.16. LOCALLY PREGALOIS CATEGORIES.

2.16.0. The amalgamation property.

A category satisfies the amalgamation property if, for any pair of monomorphisms with same source $(f : A \to B,\ g : A \to C)$, there exists a pair of monomorphisms $(m : B \to D,\ n : C \to D)$ satisfying $mf = ng$.

2.16.0.0. Proposition. For a locally simple category $\mathbb{A}$, the following assertions are equivalent :

(1) The category $\mathbb{A}$ satisfies the amalgamation property.

(2) The category $\mathbb{S}im\,\mathbb{A}$ of simple objects of $\mathbb{A}$ satisfies the amalgamation property.

(3) The monomorphisms of $\mathbb{A}$ are couniversal.

Proof : (1) => (2). If $f : A \to B$ and $g : A \to C$ are two monomorphisms of simple objects of $\mathbb{A}$, there exists two monomorphisms $m : B \to D$ and $n : C \to D$ satisfying $mf = ng$. The object D is not final, otherwise m would be a regular epimorphism, therefore an isomorphism and B would not be simple. The object D has, at least, one simple component $d : D \to Q$. The two morphisms $dm : B \to Q$ and $dn : C \to Q$ are then monomorphic and satisfy $dmf = dng$.

(2) => (3). Let $f : A \to B$ be a monomorphism of $\mathbb{A}$, $g : A \to C$ a morphism of $\mathbb{A}$ and $(m : B \to M,\ n : C \to M)$, the amalgamated sum of (f,g). Let $\Phi \in Spec_{Max}(C)$ and $\Psi = Spec_{Max}(g)(\Phi)$. The morphism $\gamma_\Phi g : A \to C/\Phi$ factorises in the form $\gamma_\Phi g = g_\Phi \gamma_\Psi$. Since f is a monomorphism, the mapping $Spec_{Max}(f) : Spec_{Max}(B) \to Spec_{Max}(A)$ is surjective, hence there is an element $\theta \in Spec_{Max}(B)$ such that $f_\theta \gamma_\Psi = \gamma_\theta f$. By the amalgamation property in $\mathbb{S}im\,\mathbb{A}$ there are two morphisms of simple objects $r_\Phi : C/\Phi \to S_\Phi$ and $s_\Phi : B_\theta \to S_\Phi$ satisfying $s_\Phi f_\theta = r_\Phi g_\Phi$. The two morphisms $s_\Phi \gamma_\theta : B \to S_\Phi$ and $r_\Phi \gamma_\Phi : C \to S_\Phi$ satisfy $s_\Phi \gamma_\theta f = s_\Phi f_\theta \gamma_\Psi = r_\Phi g_\Phi \gamma_\Psi = r_\Phi \gamma_\Phi g$. Hence, there is a morphism $\mu_\Phi : M \to S_\Phi$ satisfying $\mu_\Phi m = s_\Phi \gamma_\theta$ and $\mu_\Phi n = r_\Phi \gamma_\Phi$. Let us write

$\gamma = (\gamma_\Phi) : C \to \prod_{\Phi \in Spec_{Max}(C)} C/\Phi$, $\mu = (\mu_\Phi) : M \to \prod_{\Phi \in Spec_{Max}(C)} S_\Phi$

and $r = \prod_{\Phi \in Spec_{Max}(C)} r_\Phi : \prod_{\Phi \in Spec_{Max}(C)} C/\Phi \to \prod_{\Phi \in Spec_{Max}(C)} S_\Phi$,

then we have $\mu n = r\gamma$. However γ and r are monomorphisms, so n is also a monomorphism. Monomorphisms are thus couniversal in $\mathbb{A}$.

<u>(3) => (1)</u>. If $(f : A \to B,\ g : A \to C)$ is a pair of monomorphisms with the same source, their amalgamated sum $(m : B \to D,\ n : C \to D)$ is a pair of monomorphisms satisfying $mf = ng$. ■

2.16.0.1. <u>Examples</u>. The amalgamation property is satisfied in the categories $\mathbb{R}$ngcReg, $\mathbb{R}$ngcBa, $\mathbb{R}$ngcRegStLat, $\mathbb{R}$ngcBaStLat, $\mathbb{A}$lgcReg(k), $\mathbb{A}$lgcAlgReg(k), $\mathbb{A}$lgcAlgSep(k), $\mathbb{B}$ool.

2.16.1. <u>Objects of finite power</u>.

2.16.1.0. <u>Definition</u>. An object A in a locally finitely presentable category $\mathbb{A}$, is said to be <u>of finite power</u> if, for each finitely presentable object B of $\mathbb{A}$, the set $Hom_{\mathbb{A}}(B,A)$ is finite.

2.16.1.1. <u>Examples</u>. In categories which are algebraic over a base set, the objects of finite power are precisely the finite algebras. This is not so in categories which are algebraic over an infinite family of base sets. In the category $\mathbb{A}$lgcAlgReg(k), the algebraic field extensions are objects of finite power, and this property corresponds to the fact that a non zero polynomial $P \in k[X]$ has only a finite number of zeros in any extension field of k.

2.16.1.2. <u>Proposition</u>. <u>If A is an object of finite power in a locally finitely presentable category $\mathbb{A}$, the topological spaces $\mathbb{H}om_{\mathbb{A}}(B,A)$ are Boolean and define a functor</u> $\mathbb{H}om_{\mathbb{A}}(-,A) : \mathbb{A}^{op} \to \mathbb{E}$spBool, <u>which has a left adjoint</u>, $Exp_A : \mathbb{E}spBool \to \mathbb{A}^{op}$ <u>called the Boolean exponential functor with base A</u>.

<u>Proof</u> : If B is a finitely presentable object of $\mathbb{A}$, the topological space $\mathbb{H}om_{\mathbb{A}}(B,A)$ is finite, and thus Boolean. Any object B of $\mathbb{A}$ is a filtered colimit of the finitely presentable objects over it : say $B = \varinjlim_{i \in \mathbb{I}} B_i$. From the definition of the topology on $\mathbb{H}om_{\mathbb{A}}(B,A)$, one has $\mathbb{H}om_{\mathbb{A}}(B,A) = \varprojlim_{i \in \mathbb{I}} \mathbb{H}om_{\mathbb{A}}(B_i,A)$, and it follows that the space $\mathbb{H}om_{\mathbb{A}}(B,A)$ is Boolean. Let us denote by $E_A : \mathbb{E}nsfin \to \mathbb{A}^{op}$ the functor defined by $E_A(n) = A^n$ and, for $f : n \to p$ in $\mathbb{E}$nsfin, by

$E_{\mathbb{A}}(f) = A^f : A^p \to A^n$ where $A^f = (p_{f(i)})_{i\in[1,n]}$ and $p_i : A^n \to A$ is the ith canonical projection. Let $J : \mathbb{E}\mathrm{nsfin} \to \mathbb{E}\mathrm{spBool}$ denote the inclusion functor. It is fully faithful and codense by cofiltered limits [8, definition 2.0.]. The category $\mathbb{A}^{op}$ being complete, the right Kan extension of the functor E_A along J exists [8, Theorem 2.1.] and preserves cofiltered limits. This functor will be the Boolean exponential functor of base A and will be denoted Exp_A. Let us show that it is left adjoint to the functor $\mathbb{H}\mathrm{om}_{\mathbb{A}}(-,A)$. Consider a Boolean space X. It is a cofiltered limit of finite, hence discrete, spaces : say $X = \varprojlim_{i\in\mathbb{I}} n_i$. Let B be a finitely presentable object of $\mathbb{A}$. The space $\mathbb{H}\mathrm{om}_{\mathbb{A}}(B,A)$ is finite, hence is a finitely copresentable object in $\mathbb{E}\mathrm{spBool}$, since the finite Boolean algebras are finitely presentable in $\mathbb{B}\mathrm{ool}$. It follows that there are isomorphisms : $\mathrm{Hom}_{\mathbb{E}\mathrm{spBool}}(X,\mathbb{H}\mathrm{om}_{\mathbb{A}}(B,A)) \simeq \mathrm{Hom}_{\mathbb{E}\mathrm{spBool}}(\varprojlim_{i\in\mathbb{I}} n_i, \mathbb{H}\mathrm{om}_{\mathbb{A}}(B,A)) \simeq \varinjlim_{i\in\mathbb{I}} \mathrm{Hom}_{\mathbb{E}\mathrm{spBool}}(n_i,\mathbb{H}\mathrm{om}_{\mathbb{A}}(B,A)) \simeq \varinjlim_{i\in\mathbb{I}} \mathrm{Hom}_{\mathbb{E}\mathrm{spfin}}(n_i, \mathbb{H}\mathrm{om}_{A}(B,A)) \simeq \varinjlim_{i\in\mathbb{I}} \mathrm{Hom}_{\mathbb{A}}(B,A^{n_i}) \simeq$

$\mathrm{Hom}_{\mathbb{A}}(B, \varinjlim_{i\in\mathbb{I}} A^{n_i}) = \mathrm{Hom}_{\mathbb{A}}(B, \mathrm{Exp}_A(X))$. Any object B of $\mathbb{A}$ is a filtered colimit of the finitely presentable objects over it : say $B = \varinjlim_{k\in\mathbb{K}} B_k$. Then $\mathrm{Hom}_{\mathbb{E}\mathrm{spBool}}(X, \mathbb{H}\mathrm{om}_{\mathbb{A}}(B,A)) \simeq$

$\mathrm{Hom}_{\mathbb{E}\mathrm{spBool}}(X, \varprojlim_{k\in\mathbb{K}} \mathbb{H}\mathrm{om}_{\mathbb{A}}(B_k,A)) \simeq \varprojlim_{k\in\mathbb{K}} \mathrm{Hom}_{\mathbb{E}\mathrm{spBool}}(X, \mathbb{H}\mathrm{om}_{\mathbb{A}}(B_k,A)) \simeq$

$\varprojlim_{k\in\mathbb{K}} \mathrm{Hom}_{\mathbb{A}}(B_k, \mathrm{Exp}_A(X)) \simeq \mathrm{Hom}_{\mathbb{A}}(\varinjlim_{k\in\mathbb{K}} B_k, \mathrm{Exp}_A(X)) \simeq \mathrm{Hom}_{\mathbb{A}}(B,\mathrm{Exp}_A(X))$. ■

2.16.1.3. <u>Boolean powers of an object</u>.

The object $\mathrm{Exp}_A(X)$ is called the <u>exponential of X with base A</u> or <u>the Boolean power object of</u> A by X and is denoted by A^X. This terminology and notation extend to the morphisms in $\mathbb{E}\mathrm{spBool}$. The universal property of A^X is expressed by the isomorphism $\mathrm{Hom}_{\mathbb{A}}(-,A^X) \simeq \mathrm{Hom}_{\mathbb{E}\mathrm{spBool}}(X, \mathbb{H}\mathrm{om}_{\mathbb{A}}(-,A))$. In particular, to the morphism $1_{A^X} : A^X \to A^X$, there corresponds a continuous function $p : X \to \mathbb{H}\mathrm{om}_{\mathbb{A}}(A^X,A)$ whose values $p_x : A^X \to A$, called <u>canonical projections</u>, form a <u>continuous family $(p_x)_{x\in X}$ of morphisms</u> from A^X to A indexed by X. This family satisfies the following universal property : Given any object B of $\mathbb{A}$, there is a bijection : $\mathrm{Hom}_{\mathbb{A}}(B,A^X) \to \mathrm{Hom}_{\mathbb{E}\mathrm{spBool}}(X, \mathbb{H}\mathrm{om}_{\mathbb{A}}(B,A))$ which, to $f : B \to A^X$, associates the continuous family $(fp_x)_{x\in X}$ of morphisms from B to A, indexed by X ; in other words : for any continuous family $(f_x)_{x\in X}$ of morphisms from B to A indexed by X, there exists a unique morphism $f : B \to A^X$ satisfying $f_x = fp_x$ for each

$x \in X$. In particular, the <u>diagonal morphism</u> $\Delta_X : A \to A^X$ is defined by $p_x \Delta_X = 1_A$ for all $x \in X$.

2.16.1.4. <u>Proposition</u>. <u>The Galois group of an object of finite power in a locally finitely presentable category is a profinite group</u>.

<u>Proof</u> : Let A be an object of finite power. By Proposition 2.14.1.2., it is sufficient to prove that G_A is compact, which, by Proposition 2.16.1.2., reduces to showing that G_A is closed in $\mathbb{H}om_{\mathbb{A}}(A,A)$. Let $f : A \to A$ be a morphism which is not in G_A. There is a finitely presentable object B of A such that the function $Hom_{\mathbb{A}}(B,f) : Hom_{\mathbb{A}}(B,A) \to Hom_{\mathbb{A}}(B,A)$ is not a bijection, hence it is not injective since $Hom_{\mathbb{A}}(B,A)$ is a finite set. Thus there are two distinct morphisms $u,v : B \rightrightarrows A$ such that $fu = fv$. Then $H_{u,fu} \cap H_{v,fv}$ is an open set of $\mathbb{H}om_{\mathbb{A}}(A,A)$ containing f, but not meeting G_A. The group G_A is thus a closed subset of $\mathbb{H}om_{\mathbb{A}}(A,A)$. ■

2.16.2. <u>Locally preGalois categories</u>.

2.16.2.0. <u>Definition</u>. A category is <u>locally preGalois</u> if

(1) it is locally simple,
(2) its initial object is simple,
(3) its monomorphisms are couniversal,
(4) its simple objects are of finite power.

2.16.2.1. <u>Examples</u>. $\mathbb{A}$lgcAlgReg(k) : The category is locally simple by 2.13.1.1. Its simple objects are the algebraic extension fields of k. They are of finite power (2.16.1.1). The initial object k is simple. The amalgamation property for algebraic extension fields implies that monomorphisms are couniversal (Proposition 2.16.0.0.). The category is thus locally preGalois. One shows similarly that the categories $\mathbb{A}$lgcBaIntEng(R), $\mathbb{A}$lgcBaStLatIntEng(R), $\mathbb{A}$lgcRegStLatAlgEng(k), (examples 2.13.1.) are locally preGalois.

2.16.2.2. <u>Proposition</u>. <u>In a locally preGalois category, any endomorphism of a simple object is an automorphism</u>.

<u>Proof</u> : If f is an endomorphism of a simple object K, then it is a monomorphism. Hence for any finitely presentable object B, the mapping $Hom(B,f) : Hom(B,K) \to Hom(B,K)$ is injective and, as the set $Hom(B,K)$ is finite, will thus be bijective. Consequently f is an automorphism. ■

2.16.2.3. Theorem. A locally preGalois category has a maximum simple object i.e. a simple object L such that any simple object of the category is a subobject of L.

Proof : Let $\mathbb{A}$ be a locally preGalois category. Let $\mathcal{P}$ be a set of representatives for the finitely presentable objects of $\mathbb{A}$. Set $B = \coprod_{A\in\mathcal{P}} \coprod_{\mathbb{N}} A$. Since any simple object K of $\mathbb{A}$ is a colimit of the finitely presentable objects over it, K is a quotient object of $\coprod_{A\in\mathcal{P}} \coprod_{Hom_{\mathbb{A}}(A,K)} A$ and hence, is a quotient object of B. However, the quotient objects of B form a set [16]. Thus there is a set $\mathcal{S}$ of representatives of isomorphism classes of simple objects of $\mathbb{A}$. Let S be the sum of all the simple objects of $\mathcal{S}$. Let us show that the object S is not final. As the initial object Z is simple, the morphism $Z \to A$, for any non final object A of $\mathbb{A}$, is a monomorphism and hence, for any object B of $\mathbb{A}$, the direct image of the morphism $Z \to A$ along the morphism $Z \to B$ is a monomorphism, that is to say that the injection $B \to A \coprod B$ is a monomorphism. As a result, given any pair (A,B) of non final objects of $\mathbb{A}$, the sum $A \coprod B$ is a non final object. In particular, the sum of two simple objects of $\mathbb{A}$ is not a final object of $\mathbb{A}$ and by induction on n, the sum of n simple objects of $\mathbb{A}$ is not a final object. The property extends to infinite sums of simple objects as these are filtered colimits of finite sums and the final object is finitely presentable. The object S is thus non final. It therefore has at least one simple component, which gives a maximum simple object of $\mathbb{A}$. ■

A maximum simple object of $\mathbb{A}$ will be denoted $L_{\mathbb{A}}$.

2.16.2.4. Example. In the category $\mathbb{A}$lgcAlgReg(k), the maximum simple object is the algebraic closure of the field k.

2.16.2.5. Proposition. A locally preGalois category is locally Noetherian and its objects of finite power are precisely its semi-simple objects.

Proof : Let $\mathbb{A}$ be a locally preGalois category having $L_{\mathbb{A}}$ as its maximum simple object. Let A be a finitely presentable object of $\mathbb{A}$. Any morphism from A to $L_{\mathbb{A}}$ factorises uniquely through some simple component $A \to A/R$ with $R \in Spec_{Max}(A)$. Thus $Hom_{\mathbb{A}}(A,L_A) \simeq \coprod_{R\in Spec_{Max}(A)} Hom_{\mathbb{A}}(A/R,L_{\mathbb{A}})$. Since the set $Hom_{\mathbb{A}}(A,L_{\mathbb{A}})$ is finite and the sets $Hom_{\mathbb{A}}(A/R, L_{\mathbb{A}})$ are not empty, the set $Spec_{Max}(A)$ must be finite. Therefore the space $Spec_{Max}(A)$ is discrete and the object A, which is, the object of global sections of the structural sheaf $\tilde{A}$ on $Spec_{Max}(A)$, (Theorem 2.7.0), is a finite product of its simple components, thus is a semi-simple object. The finitely presentable objects of $\mathbb{A}$ are thus semi-simple. By Proposition 2.15.1., it follows that the category $\mathbb{A}$ is locally Noetherian. If A is an object of finite power, the Boolean algebra $\Delta(A) \simeq Hom_{\mathbb{A}}(Z^2, A)$, (1.3.1.), is finite, so the space $Spec_{Max}(A)$ is finite and the object A is semi-simple as above. Conversely, the semi-simple objects are of finite power being finite products of objects of finite power. ■

2.16.2.6. Theorem. <u>If $\mathbb{A}$ is a locally preGalois category and G is a profinite group, the category $\mathbb{A}^{Cont(G)}$ of continuous representations of G in $\mathbb{A}$ is a locally preGalois category</u>.

<u>Proof</u> : By Theorem 2.14.3.0. the category $\mathbb{A}^{Cont(G)}$ is locally simple. The initial object Z of $\mathbb{A}$ is the initial object of $\mathbb{A}^{Cont(G)}$ and it is simple in $\mathbb{A}^{Cont(G)}$ as it was so in $\mathbb{A}$. The forgetful functor $U^G : \mathbb{A}^{Cont(G)} \to \mathbb{A}$ has a right adjoint, denoted $D^G : \mathbb{A} \to \mathbb{A}^{Cont(G)}$, and hence U^G preserves amalgamated sums. It also preserves equalisers (cf. proof of Proposition 2.14.2.1.) and finite products, and thus all finite limits and monomorphisms, and it reflects isomorphisms. It follows immediately that monomorphisms are couniversal in $\mathbb{A}^{Cont(G)}$, as they were in $\mathbb{A}$. For any object (A,α) of $\mathbb{A}^{Cont(G)}$, one has the relations : $Hom_{\mathbb{A}^{Cont(G)}}((A,\alpha), D^G(L_{\mathbb{A}})) \simeq Hom_{\mathbb{A}}(U^G(A,\alpha), L_{\mathbb{A}}) = Hom_{\mathbb{A}}(A, L_{\mathbb{A}})$. It follows that, for any finitely presentable object (A,α) of $\mathbb{A}^{Cont(G)}$, there are only finitely many morphisms from (A,α) to $D^G(L_{\mathbb{A}})$. Moreover, if (K,k) is a simple object of $\mathbb{A}^{Cont(G)}$, the object K is not final in $\mathbb{A}$, hence there is a morphism $K \to L_{\mathbb{A}}$ and, thus a morphism $(K,k) \to D^G(L_{\mathbb{A}})$, which is necessarily a monomorphism. As a result, given any finitely presentable object (A,α) and any simple (K,k) in $\mathbb{A}^{Cont(G)}$, there are only finitely many morphisms from (A,α) to (K,k). The simple objects of $\mathbb{A}^{Cont(G)}$ are thus of finite power. This completes the proof that the category $\mathbb{A}^{Cont(G)}$ is a locally preGalois category. ■

2.16.2.7. <u>Definition</u> [4]. <u>A simple object</u> N in a locally preGalois category $\mathbb{A}$ is said to be <u>normal</u> if, for any pair of morphisms $(f,g) : N \rightrightarrows L_{\mathbb{A}}$, there is an automorphism σ of N satisfying $g\sigma = f$.

Let us denote by $\mathbb{A}/\!/N$ the full subcategory of $\mathbb{A}$ with objects, <u>the objects of $\mathbb{A}$ which are locally over N</u>, i.e. those objects whose simple components are subobjects of N.

2.16.2.8. <u>Theorem</u>. <u>If $\mathbb{A}$ is a locally preGalois category and N is a normal simple object of $\mathbb{A}$, the category $\mathbb{A}/\!/N$ of objects of $\mathbb{A}$ which are locally over N, is a locally preGalois category</u>.

<u>Proof</u> : Denote by $\mathbb{S}$ the full subcategory of $\mathbb{S}\mathrm{im}\,\mathbb{A}$ whose objects are subobjects of N. The category $\mathbb{L}\mathrm{oc}\,\mathbb{S}$ of objects of $\mathbb{A}$ locally belonging to $\mathbb{S}$ (2.10.0.0) is precisely the category $\mathbb{A}/\!/N$. Let us show that the category $\mathbb{S}$ is closed, in $\mathbb{S}\mathrm{im}\,\mathbb{A}$, for ultraproducts. Let K be a simple object of $\mathbb{A}$ and $\mathcal{U}$ an ultrafilter on a set I. For each set $I_o \in \mathcal{U}$, denote by $\Delta_K^{I_o} : K \to K^{I_o}$, the diagonal morphism. Denote by $K^{\mathcal{U}}$, the ultrapower of K over $\mathcal{U}$ and $\Delta_K^{\mathcal{U}} : K \to K^{\mathcal{U}}$, the diagonal morphism. For $K = L_{\mathbb{A}}$, the morphism $\Delta_{L_{\mathbb{A}}}^{\mathcal{U}} : L_{\mathbb{A}} \to L_{\mathbb{A}}^{\mathcal{U}}$ is necessarily an isomorphism. For an arbitrary simple object K, there is a monomorphism $f : K \to L_{\mathbb{A}}$ and for each $I_o \in \mathcal{U}$, the pair $(\Delta_K^{I_o} : K \to K^{I_o},\ f : K \to L_{\mathbb{A}})$ is the fibred product of the pair $(f^{I_o} : K^{I_o} \to L_{\mathbb{A}}^{I_o},\ \Delta_{L_{\mathbb{A}}}^{I_o} : L_{\mathbb{A}} \to L_{\mathbb{A}}^{I_o})$ and by passage to the filtered colimit over $\mathcal{U}$, the pair $(\Delta_K^{\mathcal{U}} : K \to K^{\mathcal{U}},\ f : K \to L_{\mathbb{A}})$ is the fibred product of the pair $(f^{\mathcal{U}} : K^{\mathcal{U}} \to L_{\mathbb{A}}^{\mathcal{U}},\ \Delta_{L_{\mathbb{A}}}^{\mathcal{U}} : L_{\mathbb{A}} \to L_{\mathbb{A}}^{\mathcal{U}})$. It follows that the morphism $\Delta_K^{\mathcal{U}}$ is an isomorphism and hence $K^{\mathcal{U}} \simeq K$. Thus, in particular, there is an isomorphism $N^{\mathcal{U}} \simeq N$. Let $(K_i)_{i\in I}$ be a family of objects of $\mathbb{S}$. For each $i \in I$, there is a monomorphism $f_i : K_i \to N$ and thus, for each $I_o \in \mathcal{U}$, a monomorphism $\prod_{i\in I_o} f_i : \prod_{i\in I_o} K_i \to N^{I_o}$, and by passage to the colimit over $\mathcal{U}$, a monomorphism $\prod_{i\in\mathcal{U}} f_i : \prod_{i\in\mathcal{U}} K_i \to N^{\mathcal{U}} \simeq N$. The ultraproduct $\prod_{i\in\mathcal{U}} K_i$ of $(K_i)_{i\in I}$ over $\mathcal{U}$ is thus an object of $\mathbb{S}$. It follows that the category $\mathbb{S}$ is closed in $\mathbb{S}\mathrm{im}\,\mathbb{A}$ under ultraproducts. By Theorem 2.10.1.0., we can conclude that the category $\mathbb{A}/\!/N = \mathbb{L}\mathrm{oc}\,\mathbb{S}$ is a variety of $\mathbb{A}$, and thus a locally simple category. Let us show that $\mathbb{A}/\!/N$ is a coreflexive subcategory of $\mathbb{A}$. Denote by $n : N \to L_{\mathbb{A}}$, a monomorphism from N into $L_{\mathbb{A}}$. Let K be a simple object of $\mathbb{A}$. There exists a monomorphism $k : K \to L_{\mathbb{A}}$. Denote by $(d : D(K) \to K,\ r : D(K) \to N)$ the fibred product of (k,n). Let M be an object of

$\mathbb{A}/\!/N$ and $g : M \to K$. There exist a monomorphism $m : M \to N$ and an automorphism σ of $L_{\mathbb{A}}$ satisfying $\sigma nm = kg$, and an automorphism μ of N satisfying $n\mu = \sigma n$. Then $n\mu m = \sigma nm = kg$. Thus there is a morphism $s : M \to D(K)$ satisfying $ds = g$ and $rs = \mu m$. The monomorphism d is thus universal from $\mathbb{A}/\!/N$ to K. This allows us to define a functor $D : \mathcal{S}im\,\mathbb{A} \to \mathbb{A}/\!/N$, partially right adjoint to the inclusion functor of $\mathbb{A}/\!/N$ into $\mathbb{A}$. As any object of $\mathbb{A}$ is an equaliser of a pair of morphisms between products of simple objects from $\mathbb{A}$, one can extend the functor D by products and equalisers to a functor $\bar{D} : \mathbb{A} \to \mathbb{A}/\!/N$ right adjoint to the inclusion functor of $\mathbb{A}/\!/N$ into $\mathbb{A}$ and consequently, $\mathbb{A}/\!/N$ is a coreflexive subcategory of $\mathbb{A}$. As a result the subcategory $\mathbb{A}/\!/N$ is closed in $\mathbb{A}$ under all limits and colimits. The initial object Z of $\mathbb{A}$ is the initial object of $\mathbb{A}/\!/N$ and is simple in $\mathbb{A}/\!/N$, as it was in $\mathbb{A}$. Monomorphisms in $\mathbb{A}/\!/N$ are couniversal as they were in $\mathbb{A}$. Let K be a simple object of $\mathbb{A}/\!/N$. Then K is a simple object in $\mathbb{A}$ thus, for any finitely presentable object A of $\mathbb{A}$, the set $\mathrm{Hom}_{\mathbb{A}}(A,K)$ is finite. If one denotes by $R : \mathbb{A} \to \mathbb{A}/\!/N$, the left adjoint to the inclusion functor, the sets $\mathrm{Hom}_{\mathbb{A}/\!/N}(RA,K) \simeq \mathrm{Hom}_{\mathbb{A}}(A,K)$ are thus finite for all finitely presentable objects A of $\mathbb{A}$. Since any finitely presentable object B of $\mathbb{A}/\!/N$ is a finite colimit of objects of the form RA, where A is finitely presentable in $\mathbb{A}$, it follows that the sets $\mathrm{Hom}_{\mathbb{A}/\!/N}(B,K)$ are finite for all finitely presentable objects B of $\mathbb{A}/\!/N$. The category $\mathbb{A}/\!/N$ is thus locally preGalois. ■

2.16.2.9. Example. If K is a normal algebraic extension field of a commutative field k, the category $\mathbb{A}\mathrm{lgcAlgReg}(k)/\!/K$ is the locally preGalois category $\mathbb{A}\mathrm{lgcAlgReg}(k/K)$ of those regular algebraic commutative k-algebras whose localisations are fields intermediate between k and K.

2.16.3. The fundamental functor.

If $\mathbb{A}$ is a locally preGalois category the functor $\mathbb{H}\mathrm{om}_{\mathbb{A}}(-,L_{\mathbb{A}}) : \mathbb{A}^{op} \to \mathbb{E}\mathrm{spBool}$ will be called the fundamental functor of $\mathbb{A}$. It is defined up to isomorphism, as the object $L_{\mathbb{A}}$ is determined up to isomorphism.

2.16.3.0. Proposition.

The fundamental functor $\mathbb{H}\mathrm{om}_{\mathbb{A}}(-,L_{\mathbb{A}}) : \mathbb{A}^{op} \to \mathbb{E}\mathrm{spBool}$ is isomorphic to the functor $\mathrm{Spec}_{\mathrm{Max}}(- \coprod L_{\mathbb{A}})$.

Proof : Put $L = L_{\mathbb{A}}$. Let A be an object of $\mathbb{A}$. Denote by $(m_A : A \to A \coprod L,\ n_A : L \to A \coprod L)$ the sum of A and L. The set $Spec_{Max}(A \coprod L)$ is in one-to-one correspondence with the set of simple components of $A \coprod L$ (c.f. 2.6.), denoted $Q_{sim}(A \coprod L)$. We will define a mapping $\gamma_A : Hom_{\mathbb{A}}(A,L) \to Q_{sim}(A \coprod L)$ by associating to $f : A \to L$, the regular quotient object $\langle f,1_L\rangle : A \coprod L \to L$. This mapping γ_A is injective, since if two morphisms f and g from A to L satisfy $\gamma_A(f) = \gamma_A(g)$, there is an isomorphism $h : L \to L$ satisfying $h\langle f,1_L\rangle = \langle g,1_L\rangle$, thus $h = 1_L$ and $f = g$. It is also surjective since, if $q : A \coprod L \to Q$ is a regular epimorphism whose target Q is a simple object, the morphism $qn_A : L \to Q$ is necessarily an isomorphism and the regular epimorphism $(qn_A)^{-1}q : A \coprod L \to L$ has the form $(qn_A)^{-1} q = \langle (qn_A)^{-1} qm_A,\ 1_L\rangle = \gamma_A((qn_A)^{-1}qm_A)$. The mappings γ_A define a natural transformation $\gamma : Hom_{\mathbb{A}}(-,L) \to Spec_{Max}(- \coprod L)$ which is an isomorphism. If A is a finitely presentable object of $\mathbb{A}$, the topological spaces $\mathbb{H}om_{\mathbb{A}}(A,L)$ and $Spec_{Max}(A \coprod L)$ are finite and discrete, thus the function $\gamma_A : \mathbb{H}om_{\mathbb{A}}(A,L) \to Spec_{Max}(A \coprod L)$ is continuous. Since both the functors, $\mathbb{H}om_{\mathbb{A}}(-,L)$ and $Spec_{Max}(- \coprod L)$ preserve cofiltered limits, and each object of $\mathbb{A}$ is a filtered colimit of finitely presentable objects, it follows that the mappings, $\gamma_A : \mathbb{H}om_{\mathbb{A}}(A,L) \to Spec_{Max}(A \coprod L)$, are continuous for all objects A of $\mathbb{A}$. As a result, they define a natural isomorphism $\gamma : \mathbb{H}om_{\mathbb{A}}(-,L) \to Spec_{Max}(- \coprod L)$. ∎

2.16.4. The fundamental group.

The Galois group of a maximum simple object $L_{\mathbb{A}}$ of a locally preGalois category $\mathbb{A}$ will be called the fundamental group of $\mathbb{A}$ and will be denoted $G_{\mathbb{A}}$. It is defined up to isomorphism and is a profinite group (Proposition 2.16.1.4.), thus a group object in the category $\mathbb{E}$spBool. It canonically defines a monad on the category $\mathbb{E}$spBool [c.f. 33, p.137] called the fundamental monad of $\mathbb{A}$, denoted $\mathbb{T}_{\mathbb{A}} = (T_{\mathbb{A}}, \eta_{\mathbb{A}}, \mu_{\mathbb{A}})$. This monad is such that $T_{\mathbb{A}}(-) = G_{\mathbb{A}} \times (-)$, $\eta_{\mathbb{A}}$ is the product with the unity : $1 \to G_{\mathbb{A}}$, and $\mu_{\mathbb{A}}$ is the product with the multiplication : $G_{\mathbb{A}} \times G_{\mathbb{A}} \to G_{\mathbb{A}}$. One knows that the category $\mathbb{E}spBool^{\mathbb{T}_{\mathbb{A}}}$ of $\mathbb{T}_{\mathbb{A}}$-algebras is equivalent to the category $G_{\mathbb{A}}$-$\mathbb{E}$spBool of Boolean $G_{\mathbb{A}}$-spaces. By 2.14.3.1., the category $G_{\mathbb{A}}$-$\mathbb{E}$spBool is equivalent to the dual of the locally simple category $\mathbb{B}ool^{Cont(G_{\mathbb{A}})}$, and the category of cosimple objects in $G_{\mathbb{A}}$-$\mathbb{E}$spBool is isomorphic to the category $G_{\mathbb{A}}$-$\mathbb{E}$spHomSep of separated homogeneous

$G_{\mathbb{A}}$-spaces, the full subcategory of the category $\mathbb{E}ns^{G_{\mathbb{A}}}$ whose objects are the separated homogeneous $G_{\mathbb{A}}$-sets i.e. the homogeneous $G_{\mathbb{A}}$-sets in which stabilisers of elements are closed subgroups of $G_{\mathbb{A}}$.

Since the simple objects of $\mathbb{A}$ are stable under subobjects, any subobject $K \to L_{\mathbb{A}}$ of $L_{\mathbb{A}}$ determines, up to isomorphism, a simple object K of $\mathbb{A}$ and, by Theorem 2.16.2.3., any simple object M of $\mathbb{A}$ is the source of at least one subobject of $L_{\mathbb{A}}$. The ordered set of subobjects of $L_{\mathbb{A}}$, denoted $\mathrm{Sim}\,\mathbb{A}$, is thus a set of representatives of isomorphism classes of simple objects of $\mathbb{A}$ and will be called the ordered set of simple objects of $\mathbb{A}$. Analogously, one considers the ordered subset $\mathrm{Sim}\,\mathbb{A}_0$ of finitely presentable simple objects of $\mathbb{A}$. Dually, in the category $G_{\mathbb{A}}$-$\mathbb{E}$spBool, the ordered set $Q(G_{\mathbb{A}})$ of quotient objects of the $G_{\mathbb{A}}$-space, $G_{\mathbb{A}}$ is a set of representatives for the cosimple objects of $G_{\mathbb{A}}$-$\mathbb{E}$spBool. However, these are the homogeneous $G_{\mathbb{A}}$-spaces, quotients of $G_{\mathbb{A}}$ by closed subgroups, and consequently the ordered set $Q(G_{\mathbb{A}})$ is isomorphic to the ordered set of closed subgroups of $G_{\mathbb{A}}$, denoted $SF(G_{\mathbb{A}})$. Similarly, the ordered set $Q_0(G_{\mathbb{A}})$ of finitely copresented cosimple objects of $G_{\mathbb{A}}$-$\mathbb{E}$spBool is isomorphic to the ordered set $SO(G_{\mathbb{A}})$ of open subgroups of $G_{\mathbb{A}}$.

2.16.4.0. Theorem. The fundamental functor, $\mathbb{H}om_{\mathbb{A}}(-,L_{\mathbb{A}}) : \mathbb{A}^{op} \to$ $\mathbb{E}$spBool, of a locally preGalois category $\mathbb{A}$, determines an adjunction $\mathbb{A}^{op} \rightleftarrows \mathbb{E}$spBool which generates the fundamental monad, $\mathbb{T}_{\mathbb{A}}$, associated to the fundamental group $G_{\mathbb{A}}$, and the comparison functor $H_{\mathbb{A}} : \mathbb{A}^{op} \to$ $G_{\mathbb{A}}$-$\mathbb{E}$spBool determines an adjunction $\mathbb{A}^{op} \rightleftarrows G_{\mathbb{A}}$-$\mathbb{E}$spBool, which induces an adjunction $(\mathrm{Sim}\,\mathbb{A})^{op} \rightleftarrows G_{\mathbb{A}}$-$\mathbb{E}$spHomSep and a Galois correspondence $(\mathrm{Sim}\,\mathbb{A})^{op} \rightleftarrows SF(G_{\mathbb{A}})$.

Proof : Put $L = L_{\mathbb{A}}$. The fundamental functor $\mathbb{H}om_{\mathbb{A}}(-,L) : \mathbb{A}^{op} \to$ $\mathbb{E}$spBool has as left adjoint the exponential functor with base L (Proposition 2.16.1.2.). It generates on the category $\mathbb{E}$spBool, the monad $\mathbb{T} = (T,\eta,\mu)$ where $T = \mathbb{H}om_{\mathbb{A}}(-,L)\ \mathrm{Exp}_L = \mathbb{H}om_{\mathbb{A}}(L^{(-)},L)$, $\eta : 1 \to T$ is defined by $\eta_X(x) = p_x$ = canonical projection of index x, (2.16.1.3), and $\mu : T^2 \to T$ is defined by $\mu_X = \mathbb{H}om_{\mathbb{A}}(\varepsilon_{L^X},L)$ where $\varepsilon : \mathrm{Exp}_L\ \mathbb{H}om_{\mathbb{A}}(-,L) \to 1_{\mathbb{A}^{op}}$ is the natural transformation whose value at A, $\varepsilon_A : A \to L^{\mathbb{H}om_{\mathbb{A}}(A,L)}$ is defined by $p_f \varepsilon_A = f$ for each morphism $f : A \to L$. The functor $\mathbb{H}om_{\mathbb{A}}(-,L)$ preserves finite sums since the object L is simple (Proposition 1.5.1.), and it preserves cofiltered limits since it is a right adjoint. The functor Exp_L preserves finite sums since it is a left adjoint and it preserves cofiltered limits (c.f. proof of 2.16.1.2.). It follows that T preserves finite sums and cofiltered limits. Given an object X of $\mathbb{E}\mathrm{spBool}$, let us define the function $\alpha_X : G_{\mathbb{A}}\times X \to \mathbb{H}\mathrm{om}_{\mathbb{A}}(L^X,L)$ by $\alpha_X(\sigma,x) = \sigma p_x$. If X is a finite space, denoted n, the function $\alpha_n : G_{\mathbb{A}}\times n \to \mathbb{H}\mathrm{om}_{\mathbb{A}}(L^n,L)$ is continuous since, for any $(\sigma_o,i_o) \in G_{\mathbb{A}}\times n$ and any basic open set $H_{u,v}$ of $\mathbb{H}\mathrm{om}_{\mathbb{A}}(L^n,L)$ containing $\alpha_n(\sigma_o,i_o) = \sigma_o p_{i_o}$, the set $\sigma_o St(p_{i_o} u) \times \{i_o\}$ is an open set in $G_{\mathbb{A}}\times n$ containing (σ_o,i_o), and its image under α_n is contained in $H_{u,v}$. Since any Boolean space, X, is a cofiltered limit of finite spaces, say $X = \varprojlim_{i\in\mathbb{I}} n_i$, it follows that the functions $\alpha_X = \varprojlim_{i\in\mathbb{I}} \alpha_{n_i}$ are continuous. These functions α_X define a natural transformation $\alpha : T_{\mathbb{A}} \to T$ and a morphism of monads $\alpha : \mathbb{T}_A \to \mathbb{T}$ [35, definition 21.1.6.], since the relations $(\alpha\eta_{\mathbb{A}})_X(x) = \alpha_X\eta_{\mathbb{A}X}(x) = \alpha_X(1,x) = p_x = \eta_X(x)$ imply $\alpha\eta_{\mathbb{A}} = \eta_X$ and that the relations : $(\mu(\alpha T)(T_{\mathbb{A}}\alpha))_X(\sigma,\tau,x) = \mu_X\alpha_{TX}(G_{\mathbb{A}}\times\alpha_X)(\sigma,\tau,x) = \mu_X\alpha_{TX}(\sigma,\tau p_x) = \mu_X(\sigma p_{\tau p_x}) = \sigma p_{\tau p_x}\varepsilon_{L}X = \sigma\tau p_x = \alpha_X(\alpha\tau,x) = \alpha_X\mu_{\mathbb{A}X}(\sigma,\tau,x)$ imply the equality $\mu(\alpha T)(T_{\mathbb{A}}\alpha) = \alpha\mu_{\mathbb{A}}$. For the singleton space 1, the value $\alpha_1 : T_{\mathbb{A}}(1) \to \mathrm{Hom}_{\mathbb{A}}(L,L)$ is an isomorphism. However, any object in $\mathbb{E}\mathrm{spBool}$ is a cofiltered limit of finite sums of the space 1, and the two functors $T_{\mathbb{A}}$ and T preserve finite sums and cofiltered limits. Thus α is an isomorphism and the two monads $\mathbb{T}_{\mathbb{A}}$ and $\mathbb{T}$ are isomorphic. The fundamental functor $\mathbb{H}\mathrm{om}_{\mathbb{A}}(-,L)$ determines a comparison functor $H_{\mathbb{A}} : \mathbb{A}^{op} \to G_{\mathbb{A}}\text{-}\mathbb{E}\mathrm{spBool}$ [33, p.138] whose value on the object L is $G_{\mathbb{A}}$ and, for any simple object K of $\mathbb{A}$, the value is a homogeneous Boolean $G_{\mathbb{A}}$-space. Since the category $\mathbb{A}$ is complete, the functor $H_{\mathbb{A}}$ has a left adjoint $F : G_{\mathbb{A}}\text{-}\mathbb{E}\mathrm{spBool} \to \mathbb{A}^{op}$ [33, p.151] whose value on the Boolean $G_{\mathbb{A}}$-space, $G_{\mathbb{A}}$, is L and, for any closed subgroup H of $G_{\mathbb{A}}$, as the homogeneous Boolean $G_{\mathbb{A}}$-space, $G_{\mathbb{A}}/H$ is a simultaneous coequaliser of the morphisms $(-)\cdot\sigma : G_{\mathbb{A}} \to G_{\mathbb{A}}$ where $\sigma \in H$, the value $F(G_{\mathbb{A}}/H)$ is the simultaneous equaliser of the morphisms $\sigma : L \to L$ belonging to H and hence, is a simple subobject of L. The adjunction $\mathbb{A}^{op} \rightleftarrows G_{\mathbb{A}}\text{-}\mathbb{E}\mathrm{spBool}$ thus induces an adjunction $(\mathbb{S}\mathrm{im}\,\mathbb{A})^{op} \rightleftarrows G_{\mathbb{A}}\text{-}\mathbb{E}\mathrm{spBoolHom}$. The two categories $G_{\mathbb{A}}\text{-}\mathbb{E}\mathrm{spBoolHom}$ and $G_{\mathbb{A}}\text{-}\mathbb{E}\mathrm{spHomSep}$ being isomorphic, one gets an adjunction $(\mathbb{S}\mathrm{im}\,\mathbb{A})^{op} \rightleftarrows G_{\mathbb{A}}\text{-}\mathbb{E}\mathrm{spHomSep}$, then an adjunction $(\mathrm{Sim}\,\mathbb{A})^{op} \rightleftarrows SF(G_{\mathbb{A}})$ which is a Galois correspondence [33, P.93]. ∎

2.16.4.1. Example. In the case of the category $\mathbb{A} = \mathbb{A}\text{lgcAlgReg}(k)$, Sim $\mathbb{A}$ is the set ExtALg(k) of algebraic extension fields of k, $G_{\mathbb{A}}$ is the Galois group $G_{\bar{k}/k}$ of the algebraic closure $\bar{k}$ of k, and the Galois correspondence $\text{ExtAlg}(k)^{op} \rightleftarrows SF(G_{\bar{k}/k})$ is classical.

2.17 LOCALLY GALOIS CATEGORIES.

Following the terminology of [35, 5.3.3.], we will say that a category is balanced if any morphism, which is both a monomorphism and an epimorphism, is an isomorphism.

2.17.0. Definition. A category is locally Galois if it is locally preGalois, balanced, and its equalisers are couniversal.

Let us note that axiom (3) of definition 2.16.2.0. is here redundant by the following proposition.

2.17.1. Proposition. *In a locally Galois category, epimorphisms and monomorphisms are regular.*

Proof : Since the category is regular in the sense of Barr [2], any morphism f factorises in the form $f = hg$ where g is a regular epimorphism and h is a monomorphism. In particular, any epimorphism f factorises in the form $f = hg$, with g a regular epimorphism and h a monomorphic epimorphism hence an isomorphism. Consequently $f \simeq g$ is a regular epimorphism. Since equalisers are couniversal, the regular monomorphisms are as well and so the category is also coregular in the sense of Barr. An analogous argument to that above proves that the monomorphisms are regular. ∎

2.17.2. Example. $\mathbb{A}$lgcAlgSep(k).
The category is locally simple by 2.13.1.2. Its simple objects are the separable algebraic extension fields of k. They are of finite power (2.16.1.1). The initial object, k, is simple. Let $f : A \to B$ be an epimorphism in $\mathbb{A}$lgcAlgSep(k). It is an epimorphism of $\mathbb{A}$lgc(k) (2.13.1.2.). The rings A and $f(A)$ are regular and thus are dominant. Consequently the epimorphic inclusion homomorphism $f(A) \to B$ is the identity. The homomorphism f is thus surjective and is a regular epimorphism in $\mathbb{A}$lgc(k). It is also a regular epimorphism in $\mathbb{A}$lgcAlgSep(k), since the

inclusion functor of $\mathbb{A}$lgcAlgSep(k) into Alg(k) preserves kernel pairs and coequalisers. Epimorphisms are thus regular in $\mathbb{A}$lgcAlgSep(k) and, consequently, the category $\mathbb{A}$lgcAlgSep(k) is balanced. Let $(f,g) : A \rightrightarrows B$ be a pair of morphisms in $\mathbb{A}$lgcAlgSep(k) and $h : C \to A$ be their equaliser. Let $r : C \to D$ be a morphism of $\mathbb{A}$lgcAlgSep(k) Equalisers and amalgamated sums are calculated in $\mathbb{A}$lgcAlgSep(k) in the same way as in $\mathbb{A}$lgc(k) since the former category is a coreflexive subcategory of the latter and is closed under subobjects, (2.13.1.2.). The rings A,B,C,D canonically have a C-module structure and the morphisms f,g,h are C-linear. The morphisms $f \otimes_C D : A \otimes_C D \to B \otimes_C D$, $g \otimes_C D : A \otimes_C D \to B \otimes_C D$, $h \otimes_C D : C \otimes_C D \to A \otimes_C D$ make up a diagram which is the direct image by r of the diagram made up of the morphisms f,g,h. However, the ring C is regular, thus absolutely flat and hence, the functor $- \otimes_C D : \mathbb{M}\text{od}(C) \to \mathbb{M}\text{od}(C)$ is exact, thus it preserves equalisers and so, $h \otimes_C D$ is the equaliser of the pair $(f \otimes_C D, g \otimes_C D)$. This proves that the equalisers in $\mathbb{A}$lgcAlgSep(k) are couniversal. ∎

2.17.3. Theorem. If $\mathbb{A}$ is a locally Galois category and G is a profinite group, the category, $\mathbb{A}^{\text{Cont}(G)}$, of continuous representations of G in $\mathbb{A}$ is locally Galois.

Proof : This theorem follows from Theorem 2.16.2.6. and the fact that the forgetful functor $U^G : \mathbb{A}^{\text{Cont}(G)} \to \mathbb{A}$ preserves equalisers, monomorphisms, amalgamated sums and epimorphisms and, that it reflects isomorphisms. ∎

2.17.4. Example. The category $\mathbb{B}$ool is locally Galois and hence, for any profinite group G, the category $\mathbb{B}\text{ool}^{\text{Cont}(G)}$ is locally Galois.

2.17.5. Theorem. If $\mathbb{A}$ is a locally Galois category and N is a normal simple object of $\mathbb{A}$, the category, $\mathbb{A}/\!/N$, of objects of $\mathbb{A}$ locally over N, is locally Galois.

Proof : This theorem follows immediately from Theorem 2.16.2.8. and the fact that $\mathbb{A}/\!/N$ is a reflexive and coreflexive full subcategory of $\mathbb{A}$ (c.f. proof of 2.16.2.8.). ∎

2.17.6. Example. If N is a normal separable field extension of a commutative field k, the category $\mathbb{A}$lgcAlgSep(k)$/\!/$N is the locally Galois category, $\mathbb{A}$lgcAlgSep(k/N) of commutative separable k-algebras whose localisations are fields intermediate between k and N.

2.17.7. Proposition. *In a locally Galois category, the maximum simple object is a proper cogenerator.*

Proof : Any object A is the object of global sections of the structural sheaf $\tilde{A}$, thus is a subobject of the product of the stalks of $\tilde{A}$, i.e. of the simple components of A. Since each simple object is a subobject of the maximum simple object L, the object A is a subobject of a power of L. Since any subobject is regular (Proposition 2.17.1.), any object A is a regular subobject of a power of L, thus L is a proper cogenerator of the category [16]. ∎

2.17.8. Theorem : *The Galois theory of a locally Galois category* :

The fundamental functor $\mathbb{H}om_{\mathbb{A}}(-,L_{\mathbb{A}}) : \mathbb{A}^{op} \to \mathbb{E}spBool$ *of a locally Galois category* $\mathbb{A}$ *is monadic and the comparison functor* $H_{\mathbb{A}} : \mathbb{A}^{op} \to G_A\text{-}\mathbb{E}spBool$ *is an equivalence of categories, inducing a duality between* :

(0) *the category* $\mathbb{A}$ *and the category* $G_{\mathbb{A}}\text{-}\mathbb{E}spBool$,

(1) *the category* $\mathbb{A}_o$ *of finitely presentable objects of* $\mathbb{A}$ *and the category* $G_{\mathbb{A}}\text{-}\mathbb{E}nsfin$ *of continuous finite* $G_{\mathbb{A}}$*-sets*,

(2) *the category* $\mathbb{S}im\,\mathbb{A}$ *of simple objects of* $\mathbb{A}$ *and the category* $G_{\mathbb{A}}\text{-}\mathbb{E}spHomSep$ *of separated homogeneous* G_A*-sets*,

(3) *the ordered set* $Sim\,\mathbb{A}$ *of simple objects of* $\mathbb{A}$ *and the ordered set* $SF(G_{\mathbb{A}})$ *of closed subgroups of* $G_{\mathbb{A}}$,

(4) *the category* $\mathbb{S}im\,\mathbb{A}_o$ *of finitely presentable simple objects of* $\mathbb{A}$ *and the category* $G_{\mathbb{A}}\text{-}\mathbb{E}nsfinHom$ *of homogeneous finite continuous* $G_{\mathbb{A}}$*-sets*,

(5) *the ordered set* $Sim\,\mathbb{A}_o$ *of finitely presentable simple objects of* $\mathbb{A}$ *and the ordered set* $SO(G_{\mathbb{A}})$ *of open subgroups of* $G_{\mathbb{A}}$.

Proof : By Theorem 2.16.4.0. and Theorem 1 of [33, p.147], it suffices to prove that the fundamental functor is monadic. By Proposition 2.16.3.0., the fundamental functor $\mathbb{H}om_{\mathbb{A}}(-,L_{\mathbb{A}})$ is isomorphic to the functor $Spec_{Max}(- \coprod L_{\mathbb{A}})$. Since equalisers are couniversal, the functor $(-) \coprod L_{\mathbb{A}} : \mathbb{A} \to \mathbb{A}$ preserves equalisers. The functor $Spec_{Max} : \mathbb{A}^{op} \to \mathbb{E}spBool$ is isomorphic to the dual of the functor $\Delta_{\mathbb{A}} : \mathbb{A} \to \mathbb{B}ool$ (1.7.). Since the functor $\Delta_{\mathbb{A}}$ preserves equalisers (Proposition 1.8.2.), the functor $Spec_{Max}$ preserves coequalisers. It follows that the functor $\mathbb{H}om_{\mathbb{A}}(-,L_{\mathbb{A}}) \sim Spec_{Max}(- \coprod L_{\mathbb{A}})$ preserves coequalisers. As the object $L_{\mathbb{A}}$ is a proper cogenerator in $\mathbb{A}$ (Proposition 2.17.7.), the functor $\mathbb{H}om_{\mathbb{A}}(-,L_{\mathbb{A}})$ reflects isomorphisms. It then follows from Beck's theorem characterising monadic functors [33, p.147] that the functor $\mathbb{H}om_{\mathbb{A}}(-,L_{\mathbb{A}})$ is monadic. ■

2.17.9. Remark on the different Galois theories.
Each of the dualities annonced in Theorem 2.17.8. is a "Galois theory". Duality (5) is the classical Galois theory and duality (3) is its extension to extensions of infinite degrees. Duality (4) is the "abstract" Galois theory described by M. Barr [4] and duality (2) is its extension to extensions of infinite degrees. Duality (1) is the "axiomatic" Galois theory described by A. Grothendieck [20] and duality (0) is its extension to extensions of infinite degrees. To obtain the Galois theory of fields, one considers the locally Galois category $\mathbb{A}$lgcAlgSep(k).

2.17.10. Theorem. Locally Galois categories are precisely categories equivalent to the categories $\mathbb{B}ool^{Cont(G)}$ of continuous representations of a profinite group G in Boolean algebras.

Proof : The categories $\mathbb{B}ool^{Cont(G)}$ are locally Galois (example 2.17.4.) and, by Theorem 2.17.8., any locally Galois category $\mathbb{A}$ is equivalent to the category $G_{\mathbb{A}}$-$\mathbb{E}spBool^{op}$, which is itself equivalent to the category $\mathbb{B}ool^{Cont(G_{\mathbb{A}})}$ (example 2.14.3.1.). ■

CHAPTER III

CATEGORIES OF LOCALLY BOOLEAN SHEAVES OF SIMPLE ALGEBRAS

3.1. QUASI LOCALLY SIMPLE CATEGORIES.

3.1.0. Notations.

Let $\mathbb{A}$ be a locally finitely presentable category [16] and A an object of $\mathbb{A}$. Following the notation of [35, 14.2], we will denote by Mono(A), the class of monomorphisms with target A, preordered by the relation : $m : X \to A \leqslant n : Y \to A$ if, and only if, there is a morphism $f : X \to Y$ satisfying $m = nf$. If $(m_i : X_i \to A)_{i \in I}$ is a non-empty family of members of Mono(A) having a limit $(n_i : X \to X_i)_{i \in I}$ in $\mathbb{A}$, the common value $m = m_i n_i : X \to A$ is a monomorphism called the intersection of the m_i $(i \in I)$ and is denoted $\bigcap_{i \in I} m_i$; it is the meet of the m_i $(i \in I)$ in Mono(A). The join of the m_i $(i \in I)$ in Mono(A) is called the union of the m_i $(i \in I)$ and is denoted $\bigcup_{i \in I} m_i$. The ordered set associated to Mono(A) is denoted S(A). Its elements are the subobjects of A and its meets and joins are written as in Mono(A). Dually, one denotes by Epi(A), the class of epimorphisms with source A, preordered by the relation $p : A \to P \leqslant q : A \to Q$ if, and only if, there is a morphism $f : Q \to P$ satisfying $p = fq$. The meet of a family $(q_i : A \to Q_i)_{i \in I}$ of members of Epi(A) is called the cointersection of the q_i $(i \in I)$, is denoted $\bigcap_{i \in I} q_i$ and is defined by a colimit. The join of the q_i $(i \in I)$ is called the counion and is denoted $\bigcup_{i \in I} q_i$. The ordered set associated to Epi(A) is denoted Q(A) and its elements are the quotient objects of A. A family $(q_i : A \to Q_i)_{i \in I}$ of members of Epi(A) or Q(A) is called a counion if $\bigcup_{i \in I} q_i = 1_A$. If $q : A \to Q$ is a member of Epi(A) and $(q_i : Q \to Q_i)_{i \in I}$ is a family of members of Epi(Q), then $q = \bigcup_{i \in I} q_i q \Leftrightarrow 1_Q = \bigcup_{i \in I} q_i \Leftrightarrow$ the family $(q_i)_{i \in I}$ is a counion. A counion $(q_i : Q \to Q_i)_{i \in I}$ is couniversal if, for any morphism $f : Q \to P$, the family $(p_i : P \to P_i)_{i \in I}$ of direct images, i.e. obtained by amalgamated sums, of the morphisms q_i by f, is a counion. A regular epimorphism is a morphism $q : A \to Q$ which is a coequaliser of a pair of morphisms $(m,n) : X \rightrightarrows A$ [2]. It represents a regular quotient object of A. It is said to be finitely generated if X is a finitely generated object of

$\mathbb{A}$ or, what amounts to the same thing, if X is a finitely presentable object of $\mathbb{A}$ [16]. A <u>null object</u> of $\mathbb{A}$ is an object which is both initial and final ; it will be denoted 0 as will be all the morphisms of the form $A \to 0 \to B$. For a morphism $f : A \to B$, the equaliser of the pair, $(f,0)$, is the <u>kernel</u> of f, and its coequaliser is the <u>cokernel</u> of f. A product $(p_i : A \to A_i)_{i \in I}$ of objects of $\mathbb{A}$ is codisjoint if the canonical projection p_i are epimorphisms and $p_i \cap p_j = 0$ for $i \neq j$.

3.1.1. <u>Definition</u>. A category is <u>quasi locally simple</u> if

(1) it is locally finitely presentable and has a null object,

(2) its products are codisjoint,

(3) its counions of pairs of regular quotient objects are couniversal,

(4) its finitely generated regular quotient objects are complemented and are coclassified by an object Z.

The <u>coclassifier</u>, Z, satisfies the following property : there is a finitely generated regular epimorphism $q : Z \to Q$ such that, for any finitely generated regular epimorphism $p : A \to P$, there is a unique morphism $e : Z \to A$ such that the direct image of q by e is p. In particular, there must exist a unique morphism, $e : Z \to Q$ such that the direct image of q by e is 1_Q. However the morphisms $e = q$ and $e = 0$, both make this work, so $q = 0$ and, as q is an epimorphism, $Q = 0$. It follows that any finitely generated regular epimorphism $p : A \to P$ is the cokernel of a unique morphism $e : Z \to A$.

3.1.2. <u>Examples</u>. Contrary to our previous custom, we will consider here not necessarily unitary rings and, in the categories which follow, called categories of non-unitary rings, and denoted by the symbol Nu, the objects may be unitary rings, or non-unitary rings. The corresponding categories of unitary rings have been studied in chapter 2.

3.1.2.0. <u>$\mathbb{R}$ngStRegNu</u> : The category of strongly regular non-unitary rings.

This is an algebraic category in which the ring $\{0\}$ is null. Since the rings are non-empty sets, their products are codisjoint. The regular quotients of an object A are its quotients by ideals. If I and J are two ideals of A, the counion of the pair $(A \to A/I, A \to A/J)$

is the quotient $A \to A/I \cap J$. It follows that $(A \to A/I, A \to A/J)$ is a counion if, and only if, $I \cap J = \{0\}$. Then, its direct image by a morphism $f : A \to B$ is the pair $(B \to B/Bf(I), B \to B/Bf(J))$ whose counion is the quotient $B \to B/Bf(I) \cap Bf(J)$. However an element b of $Bf(I) \cap Bf(J)$ is of the form $b = b_1 f(x_1) + \ldots + b_n f(x_n) = c_1 f(y_1) + \ldots + c_p f(y_p)$ with $x_1, \ldots, x_n \in I$ and $y_1, \ldots, y_p \in J$. Its square b^2 is of the form $\Sigma f(xy)$ with $x \in I$ and $y \in J$, hence it is null. Since the ring B is reduced, $b = 0$, hence $Bf(I) \cap Bf(J) = \{0\}$ and the pair $(B \to B/Bf(I), B \to B/Bf(J))$ is a counion. Counions of pairs of regular epimorphisms are thus couniversal. Every finitely generated ideal of A is generated by an idempotent, e. Then $e^{\perp} = \{x : xe = 0\}$ is an ideal of A such that $Ae \cap e^{\perp} = 0$ and $Ae + e^{\perp} = A$. It follows that the pair $(A \to A/Ae, A \to A/e^{\perp})$ is a product and that the quotient object $A \to A/e$ is complemented. The strongly regular unitary ring universally associated to the ring $\mathbb{Z}$ is the idempotent coclassifier, and hence, is coclassifier for finitely generated regular quotients.

3.1.2.1. <u>$\mathbb{R}$ngcRegNu</u> : non-unitary commutative regular rings.

3.1.2.2. <u>$\mathbb{R}$ngcRegNu(p</u>) : non-unitary commutative regular rings of prime characteristic p.

3.1.2.3. <u>$\mathbb{R}$ngcNu(p)</u> : non-unitary p-rings

3.1.2.4. <u>$\mathbb{R}$ngcNu(p^k</u>) : non-unitary p^k-rings.

3.1.2.5. <u>$\mathbb{R}$ngcRegStLatNu</u> : non-unitary strongly lattice-ordered commutative regular rings (i.e. non-unitary regular f-rings).

3.1.2.6. <u>$\mathbb{R}$ngcBaNu</u> : non-unitary commutative Baer rings.

A unitary commutative Baer ring is defined as being a commutative ring in which the annihilator of each element is an ideal generated by an idempotent. This comes to the same thing as associating to each element x, an idempotent $e(x)$ satisfying $xe(x) = x$ and $(yx = 0 \Rightarrow ye(x) = 0)$. This second description can be used to define non-unitary Baer rings and, with ring homomorphisms which preserve associated idempotents, they make up the above category.

3.1.2.7. <u>$\mathbb{R}$ngcBaLatNu</u> : non-unitary lattice-ordered commutative Baer rings with positive squares.

3.1.2.8. <u>$\mathbb{R}$ngcBaStLatNu</u> : non-unitary strongly lattice-ordered commutative Baer rings.

3.1.2.9. $\mathbb{R}$ngcStLatProjNu : non-unitary, projectable, strongly lattice-ordered, commutative rings.

A unitary, projectable, strongly lattice-ordered, commutative ring is defined as being a strongly lattice-ordered commutative ring in which each principal polar ideal is a direct factor. This amounts to associating to each element x, an idempotent $e(x)$ satisfying $xe(x) = x$ and $(|y| \wedge |x| = 0 \Rightarrow ye(x) = 0)$. It is with this second description that one defines non unitary projectable strongly lattice-ordered rings.

3.1.2.10. $\mathbb{R}$ng$\mathbb{E}$Nu : non-unitary E-rings.

Following Kennison, an E-ring is a ring, which has a unary operation e satisfying the axioms :

(1) $e(x)$ is a central element

(2) $e(e(x)) = e(x)$

(3) $e(0) = 0$

(4) $xe(x) = x$

(5) $e(y\, e(x)) = e(y)\, e(x)$

With those ring homomorphisms which preserve the operation e, they constitute the category $\mathbb{R}$ng$\mathbb{E}$Nu. It is an algebraic category having the ring $\{0\}$ as a null object and in which products are codisjoint. Let A be an E-ring. The idempotents of A of the form $e(x)$ are said to be stable, as are the ideals I of A such that $e(I) \subset I$. The stable ideal of A generated by an element a of A is also generated by the stable idempotent $e(a)$. The regular quotient objects of A are the quotients of A by the stable ideals of A. If I and J are two ideals of A, the counion of the pair $(A \to A/I,\ A \to A/J)$ is the quotient $A \to A/I \cap J$. Suppose that this pair is a counion i.e. $I \cap J = \{0\}$. Its direct image by a morphism $f : A \to B$ is the pair $(B \to B/Bf(I),\ B \to B/Bf(J))$, the counion of which is the quotient $B \to B/Bf(I) \cap Bf(J)$. However an element b of this intersection is of the form $b = b_1 f(x_1) + \ldots + b_n f(x_n) = c_1 f(y_1) + \ldots + c_p f(y_p)$, where $x_1, \ldots, x_n$ are some stable idempotents of I and $y_1, \ldots, y_p$, some stable idempotents of J, and if one denotes by $u \vee v = u+v-uv$, the join of two idempotents u and v, one then has : $b = (b_1 f(x_1) + \ldots + b_n f(x_n))(f(x_1) \vee \ldots \vee f(x_n)) = (c_1 f(y_1) + \ldots + c_p f(y_p))(f(x_1) \vee \ldots \vee f(x_n)) = c_1 f_1(y_1)(f(x_1) \vee \ldots \vee f(x_n)) + \ldots + c_p f(y_p)(f(x_1) \vee \ldots \vee f(x_n)) = c_1(f(y_1 x_1)) \vee \ldots \vee f(y_1 x_n)) + \ldots + c_p(f(y_p x_1) \vee \ldots \vee f(y_p x_n)) = 0$ since $y_j x_i \in I \cap J = \{0\}$. It follows that $Bf(I) \cap Bf(J) = 0$ and, consequently, that counions of pairs of regular quotient objects are couniversal. Since any stable ideal of A, generated by one element, is generated by a stable idempotent, any finitely generated stable ideal I of A is also generated by a stable idempotent, u, and $u^{\perp} = \{x \in A : xu = 0\}$ is a stable ideal of A such that $I \oplus u^{\perp} = uA \oplus u^{\perp} = A$. Consequently, the pair $(A \to A/I, A \to A/u^{\perp})$ is a product and the quotient $A \to A/I$ is complemented. The ring $\mathbb{Z}$ together with the unary operation e defined by $e(0) = 0$ and $e(n) = 1$ for $n \neq 0$, is an E-ring which is a coclassifier for stable idempotents, thus a coclassifier for finitely generated regular quotient objects.

3.1.2.11. <u>RngcENu</u> : non-unitary commutative E-rings.

3.1.2.12. <u>RngcEStLatNu</u> : non-unitary strongly lattice-ordered commutative E-rings.

3.1.2.13. <u>RngDifRegNu</u> : non-unitary regular differential commutative rings.

3.1.2.14. <u>RngDifRegNu(p)</u> : non-unitary regular differential commutative rings of prime characteristic, p.

3.1.2.15. <u>RngDifRegPerfNu(p)</u> : non-unitary regular differential rings which are differentially perfect and of prime characteristic p.

3.1.2.16. <u>BoolNu</u> : non-unitary Boolean algebras.

These are also called generalised Boolean algebras [18]. They are distributive lattices with a least element, and for which, for any element a, the sublattice $\{x : x \leqslant a\}$ is complemented, i.e. is a Boolean algebra. This notion is equivalent to that of a non-unitary Boolean ring.

3.1.2.17. <u>Any variety</u> in one of the above categories.

3.1.2.18. <u>Any category of algebras</u> whose underlying rings are coming from one of the above categories.

3.1.2.19. <u>Any finite product</u> of the above categories.

3.2. QUASI LOCALLY SIMPLE CATEGORIES OF AUGMENTED OBJECTS.

Given a category $\mathbb{X}$ with final object 1, a pointed object of $\mathbb{X}$ is a pair (X,x) where X is an object of $\mathbb{X}$ and $x : 1 \to X$ is a morphism [35, 11.1.2.], and a morphism of pointed objects $(X,x) \to (Y,y)$ is a morphism $f : X \to Y$ of $\mathbb{X}$ satisfying $fx = y$. Pointed objects in $\mathbb{X}$, and their morphisms, form the category $\mathbb{X}_*$, which is none other than the category of objects of $\mathbb{X}$ under the object 1. Dually, if Z is an initial object of $\mathbb{X}$, an augmented object (or co-pointed object) of $\mathbb{X}$ is a pair (A,a) where A is an object of $\mathbb{X}$ and $a : A \to Z$ is a morphism [33, p.46], and a morphism of augmented objects $(A,a) \to (B,b)$ is a morphism $f : A \to B$ of $\mathbb{X}$ satisfying $bf = a$. These form the category $\mathbb{X}^*$, which is none other than the category of objects of $\mathbb{X}$ over the object Z. The canonical projection functor $P : \mathbb{X}^* \to \mathbb{X}$ is defined by $P(A,a) = A$ and $Pf = f$. It is a fibred functor which creates colimits and connected limits. If $\mathbb{X}$ has finite products, the canonical injection functor, $J : \mathbb{X} \to \mathbb{X}^*$, is defined by $JA = (Z\times A, p_A)$ where $p_A : Z\times A \to Z$ is the canonical projection, and by $Jf = 1_Z\times f$. The functor P is left adjoint to J. It is comonadic and generates the comonad $\mathbb{G} = (G,\varepsilon,\nu)$, on $\mathbb{X}$ where $G=(-)\times Z$, ε is the canonical projection, and ν is the product with the diagonal of Z.

3.2.0. Theorem. If $\mathbb{A}$ is a locally simple category (2.1.0), the category $\mathbb{A}^*$ of augmented objects of $\mathbb{A}$ is quasi locally simple.

Proof : a) Let us prove that the functor $J : \mathbb{A} \to \mathbb{A}^*$ reflects isomorphisms. If $f : A \to B$ is a morphism of $\mathbb{A}$ such that $Jf = 1_Z\times f$ is an isomorphism and, if one denotes by $(p_A : Z\times A \to Z,\ q_A : Z\times A \to A)$ the product of Z and A, and $(p_B : Z\times B \to Z,\ q_B : Z\times B \to B)$ the product of Z and B, then, the pair (f,q_B) is the amalgamated sum of $(q_A, 1_Z\times f)$, since finite products are couniversal in $\mathbb{A}$. Consequently f is an isomorphism.

b) Let us show that the functor J preserves connected colimits. Since the functor $P : \mathbb{A}^* \to \mathbb{A}$ reflects colimits, it suffices to show that the functor PJ preserves connected colimits. Let $A = \varinjlim_{i\in\mathbb{I}} A_i$, be a connected colimit in $\mathbb{A}$. Denote by $(Z_i)_{i\in\mathbb{I}}$, the constant connected diagram with value Z and base $\mathbb{I}$. Then $Z = \varinjlim_{i\in\mathbb{I}} Z_i$. For each simple object, S, of $\mathbb{A}$, one has the natural isomorphisms :
$\mathrm{Hom}_{\mathbb{A}}(PJA,S) = \mathrm{Hom}_{\mathbb{A}}(Z\times A,S) \simeq \mathrm{Hom}_{\mathbb{A}}(Z,S) \coprod \mathrm{Hom}_{\mathbb{A}}(A,S) \simeq \mathrm{Hom}_{\mathbb{A}}(\varinjlim_{i\in\mathbb{I}} Z_i,S)$ $\coprod \mathrm{Hom}_{\mathbb{A}}(\varinjlim_{i\in\mathbb{I}} A_i,S) \simeq \varprojlim_{i\in\mathbb{I}} \mathrm{Hom}_{\mathbb{A}}(Z_i,S) \coprod \varprojlim_{i\in\mathbb{I}} \mathrm{Hom}_{\mathbb{A}}(A_i,S) \simeq$

$\simeq \varprojlim_{i\in\mathbb{I}}(\mathrm{Hom}_{\mathbb{A}}(Z_i,S) \coprod \mathrm{Hom}_{\mathbb{A}}(A_i,S)) \simeq \varprojlim_{i\in\mathbb{I}} \mathrm{Hom}_{\mathbb{A}}(Z_i\times A_i,S) \simeq$ $\varprojlim_{i\in\mathbb{I}} \mathrm{Hom}_{\mathbb{A}}(PJA_i,S) \simeq \mathrm{Hom}_{\mathbb{A}}(\varinjlim_{i\in\mathbb{I}} PJA_i,S)$. Since the simple objects of $\mathbb{A}$ form a proper cogenerating class for $\mathbb{A}$, it follows that there is an isomorphism $PJA \simeq \varinjlim_{i\in\mathbb{I}} PJA_i$.

c) Let us show that the functor $P : \mathbb{A}^* \to \mathbb{A}$ preserves and reflects the finitely presentable objects. Let (A,a) be a finitely presentable object of $\mathbb{A}^*$. For any filtered colimit $\varinjlim_{i\in\mathbb{I}} B_i$ in $\mathbb{A}$, one has isomorphisms : $\mathrm{Hom}_{\mathbb{A}}(A, \varinjlim_{i\in\mathbb{I}} B_i) = \mathrm{Hom}_{\mathbb{A}}(P(A,a), \varinjlim_{i\in\mathbb{I}} B_i) \simeq$ $\mathrm{Hom}_{\mathbb{A}^*}((A,a), \varinjlim_{i\in\mathbb{I}} JB_i) \simeq \varinjlim_{i\in\mathbb{I}} \mathrm{Hom}_{\mathbb{A}^*}((A,a),JB_i) \simeq \varinjlim_{i\in\mathbb{I}} \mathrm{Hom}_{\mathbb{A}}(P(A,a),B_i)$ $= \varinjlim_{i\in\mathbb{I}} \mathrm{Hom}_{\mathbb{A}}(A,B_i)$. The object A is thus finitely presentable in $\mathbb{A}$. The converse is immediate. It follows that if $\mathbb{A}_o$ denotes the category of finitely presentable objects of $\mathbb{A}$, the category of finitely presentable objects of $\mathbb{A}^*$ is precisely the category $\mathbb{A}_o^*$ of augmented objects in $\mathbb{A}_o$.

d) Let us show that the functor J preserves and reflects finitely presentable objects, and that any finitely presentable object of $\mathbb{A}^*$ is of the form JA. Let A be an object of $\mathbb{A}$. If it is finitely presentable, the object $PJA = Z\times A$ is also, as it is the product of the finitely presentable object, Z, (c.f. proof of Theorem 3.6.5.) by the finitely presentable object A (Proposition 1.4.0). Thus the object JA is finitely presentable by c) above. Conversely, if the object $JA = (Z\times A,p_A)$ is finitely presentable, the object $PJA = Z\times A$ is so as well by c), thus so is the object A, which is a finitely generated regular quotient of $Z\times A$. Let (B,b) be a finitely presentable object of $\mathbb{A}^*$. The object $B = P(B,b)$ is finitely presentable and the morphism $b : B \to Z$ is a split epimorphism, thus a coequaliser of two morphisms of source B, and it follows that b is a finitely generated regular epimorphism, thus a direct factor of the form $p_A : Z\times A \to Z$. The object (B,b) is thus of the form JA.

e) The category $\mathbb{A}^*$ is cocomplete since the category $\mathbb{A}$ is, and the functor P creates colimits. Let $f : (A,a) \to (B,b)$ be a morphism of $\mathbb{A}^*$ such that the functions $\mathrm{Hom}_{\mathbb{A}^*}((X,x),f)$ are bijective for all finitely presentable objects (X,x) of $\mathbb{A}^*$. Let X be a finitely presentable object of $\mathbb{A}$. If $m,n : X \rightrightarrows A$ satisfy $fm = fn$, then $am = bfm = bfu = an$, thus (X, am) is a finitely presentable object of $\mathbb{A}^*$ determining two morphisms $m,n : (X,am) \rightrightarrows (A,a)$ which equalise the morphism $f : (A,a) \to (B,b)$ and, consequently, one has the equality : $m = n$. The mapping $\mathrm{Hom}_{\mathbb{A}}(X,f)$ is thus injective. For any morphism $g : X \to B$, the morphism $g : (X,bg) \to (B,b)$ determines at least one morphism $h : (X,bg) \to (A,a)$ satisfying $fh = g$, that is to say, a morphism $h : X \to A$ satisfying $fh = g$. It follows that the mapping $\mathrm{Hom}_{\mathbb{A}}(X,f)$ is bijective and, hence, that the morphism $f : A \to B$ is an isomorphism, the same then being true for $f : (A,a) \to (B,b)$. It follows that the finitely presentable objects of $\mathbb{A}^*$ form a proper generating set for $\mathbb{A}^*$ and that the category $\mathbb{A}^*$ is locally finitely presentable. Moreover, $(Z,1_Z)$ is a null object in $\mathbb{A}^*$.

f) Since finite products are codisjoint in $\mathbb{A}$ and, as the functor J preserves and reflects products and amalgamated sums, it follows from d) that finite products of finitely presentable objects of $\mathbb{A}^*$ are codisjoint. Let M and N be two objects of $\mathbb{A}^*$. They are filtered colimits of finitely presentable objects of $\mathbb{A}^*$: say $M = \varinjlim_{i\in\mathbb{I}} M_i$ and $N = \varinjlim_{i\in\mathbb{I}} N_i$. If one denotes by $(p_i : M_i{\times}N_i \to M_i$, $q_i : M_i{\times}N_i \to N_i)$, the product of M_i and N_i, and by $(p : M{\times}N \to M$, $q : M{\times}N \to N)$, the product of M and N, one obtains $p = \varinjlim_{i\in\mathbb{I}} p_i$ and $q = \varinjlim_{i\in\mathbb{I}} q_i$. Then the amalgamated sum of (p,q) is the filtered colimit over $\mathbb{I}$ of the amalgamated sums of (p_i,q_i). It is thus null. Thus products are codisjoint in $\mathbb{A}^*$.

g) Let us show that, in the locally simple category $\mathbb{A}$, counions of pairs of regular quotient objects are fibred products of their amalgamated sums and are couniversal. Consider, first, a pair of direct factors $(\delta : A \to B,\ \varepsilon : A \to C)$. Let $\mu : A \to M$ be the join of δ and ε in $\Delta(A)$, $\nu : A \to N$, their meet which is the amalgamated sum of (δ,ε), and $\delta_1 : B \to N$, $\varepsilon_1 : C \to N$, $\delta_2 : M \to B$, $\varepsilon_2 : M \to C$ the morphisms which satisfy $\delta_1\delta = \nu = \varepsilon_1\varepsilon$, $\delta_2\mu = \delta$ and $\varepsilon_2\mu = \varepsilon$. The structural sheaf $\tilde{A}$ takes the values $\tilde{A}(D(\delta)) = B$, $\tilde{A}(D(\varepsilon)) = C$, $\tilde{A}(D(\delta) \cap D(\varepsilon)) = \tilde{A}(D(\nu)) = N$ and $\tilde{A}(D(\delta) \cup D(\varepsilon)) = \tilde{A}(D(\mu)) = M$. From the glueing property of the sheaf, $\tilde{A}$, it follows that the pair of morphisms $(\delta_2 : M \to B,\ \varepsilon_2 : M \to C)$ is the fibred product of the pair $(\delta_1 : B \to N,\ \varepsilon_1 : C \to N)$ and, consequently, that μ is the counion of δ and ε. This counion is couniversal since, for any morphism $f : A \to D$, the direct image $\Delta(f)(\mu)$ of μ by f is the join of the direct images $\Delta(f)(\delta)$ and $\Delta(f)(\mu)$ of δ and ε by f, (1.8.2.), thus is the counion of these images. Consider next a pair $(p : A \to P,\ q : A \to Q)$ of regular quotient objects. They are filtered cointersections (i.e. filtered colimits) of direct factors : say $p = \bigcap_{i\in I} p_i$ and $q = \bigcap_{I\in I} q_i$ (2.1.2.). For each $i \in I$, the counion μ_i of p_i and q_i is the fibred product of the amalgamated sum of (p_i, q_i) and is couniversal. By passing to the filtered colimit over I, the filtered colimit $= \varinjlim_{i\in I} \mu_i$ is the fibred product of the amalgamated sum of (p,q) and is the couniversal counion of p and q.

h) Let $(p : X \to P,\ q : X \to Q)$ be a counion of regular epimorphisms in $\mathbb{A}^*$. Let $(m : P \to S,\ n : Q \to S)$ be the amalgamated sum of (p,q) ; $(r : Y \to P,\ s : Y \to Q)$, the fibred product of (m,n) and $y : X \to Y$, the morphism satisfying $ry = p$ and $sy = q$. Since the functor P preserves regular epimorphisms, amalgamated sums and fibred products, the morphism Py is the counion of Pp and Pq. The functor P being faithful, the morphism y is an epimorphism and hence an isomorphism. The pair (p,q) is thus the fibred product of its amalgamated sum. It follows that the functor P preserves and reflects counions of regular epimorphisms. The couniversality of counions of pairs of regular epimorphisms of $\mathbb{A}^*$ then follows from that of $\mathbb{A}$.

i) Let $p : (A,a) \to (B,b)$ be a finitely generated regular epimorphism of $\mathbb{A}^*$. Its image $p : A \to B$ by the functor P is a finitely generated regular epimorphism of $\mathbb{A}$, because the functor P preserves finitely presentable objects and coequalisers. The morphism p is thus a direct factor of the form $p : B\times C \to B$ (2.1.2.). Then the pair of morphisms $(b\times 1_C : (B\times C,a) \to (Z\times C,p_C),\ p : (B\times C,a) \to (B,b))$ is the fibred product of the pair of morphisms $(p_C : (Z\times X,p_C) \to (Z,1_Z),\ b : (B,b) \to (Z,1_Z))$ because its image by P is a fibred product in $\mathbb{A}$ and the functor P reflects fibred products. Since the object $(Z,1_Z)$ is the final object of $\mathbb{A}^*$, the above pair $(b\times 1_C, p)$ is, in fact, a product in $\mathbb{A}^*$. It follows that the morphism $p : (A,a) \to (B,b)$ is a complemented regular epimorphism. Denoting by $i_B : Z \to B$, $i_C : Z \to C$ the uniquely defined morphisms and $\pi : Z^2 \to Z$ the first projection, one knows that the pair $(p : A \to B,\ i_B : Z \to B)$ is the amalgamated sum of the pair $(i_B\times i_C : Z^2 \to A,\ \pi : Z^2 \to Z)$ in $\mathbb{A}$ (Theorem 1.3.4). It follows that the pair $(p : (A,a) \to (B,b),\ i_B : (Z,1_Z) \to (B,b))$ is the amalgamated sum of the pair $(i_B\times i_C : (Z^2,\pi) \to (A,a),\ \pi : (Z^2,\pi) \to (Z,1_Z))$ since the functor P reflects amalgamated sums. Hence, the object (Z^2,π) coclassifies the finitely generated regular quotient objects in $\mathbb{A}^*$. ■

3.3. QUASI LOCALLY SIMPLE CATEGORIES OF FUNCTORS.

Let $\mathbb{C}$ be a finitely complete category having disjoint and universal finite sums and whose objects are decidable [23, p.162] i.e. such that the diagonal monomorphisms $\Delta_X : X \to X\times X$ are complemented. The category $\mathbb{P}\mathrm{art}(\mathbb{C})$ of regular partial morphisms of $\mathbb{C}$ has, as its objects, the objects of $\mathbb{C}$ and, as its morphisms from A to B, the pairs (u,f) formed from a regular subobject $u : U \to A$ of A and a morphism $f : U \to B$, the composition being defined by $(v,g)(u,f) = (uu_1, gf_1)$ where (u_1,f_1) is the fibred product of (f,v).

3.3.0. Theorem. *If $\mathbb{C}$ is a finitely complete small category with disjoint and universal finite sums and in which the objects are decidable, the category $\mathbb{C}\mathrm{ont}_{\aleph_0}[\mathbb{P}\mathrm{art}(\mathbb{C}), \mathbb{E}\mathrm{ns}]$ of finitely continuous functors defined on the category $\mathbb{P}\mathrm{art}(\mathbb{C})$ of regular partial morphisms of $\mathbb{C}$, and with values in $\mathbb{E}\mathrm{ns}$, is quasi locally simple.*

Proof : Let us first show that the category, $\mathbb{P}\mathrm{art}(\mathbb{C})$, is equivalent to the category $\mathbb{C}_*$ of pointed objects in $\mathbb{C}$. In the category $\mathbb{C}$, any diagonal monomorphism $\Delta_X : X \to X\times X$ is a direct summand and forms with its complement $\Delta'_X : X' \to X\times X$, a sum (Δ_X, Δ'_X) of X and X' (proof of Theorem 2.3.0). It follows from the universality of finite sums that any regular subobject $u : U \to A$ is complemented and forms, with its complement $u' : U' \to A$ a sum of U and U'. For each object A of $\mathbb{C}$, we will denote by $(e_A : 1 \to 1 \amalg A,\ i_A : A \to 1 \amalg A)$, the sum of the final object, 1, with A. The functor $E : \mathbb{P}\mathrm{art}(\mathbb{C}) \to \mathbb{C}_*$ is defined, for an object A of $\mathbb{C}$, by $EA = (1 \amalg A, e_A)$ and for a morphism $(u : U \to A,\ f : U \to B)$ by $E(u,f) = t_u \amalg f : 1 \amalg A = 1 \amalg U' \amalg U \to 1 \amalg B$ where $t_u : 1 \amalg U' \to 1$ denotes the uniquely defined morphism. E is a functor because $E(v,g)E(u,f) = (t_v \amalg g)(t_u \amalg f) = t_{uu_1} \amalg gf_1 = E(uu_1, gf_1) = E((v,g)(u,f))$. The functor E is faithful because, for any morphism $(u,f) : A \to B$ of $\mathbb{P}\mathrm{art}(\mathbb{C})$, the pair $(i_A u, f)$ is the fibred product of the pair $(E(u,f), 1_B)$. Let A and B be two objects of $\mathbb{C}$ and $g : E(A) \to E(B)$ a morphism of $\mathbb{C}_*$. If $(t : U_1 \to 1,\ u_1 : U_1 \to 1 \amalg A)$ is the fibred product of $(e_B : 1 \to 1 \amalg B,\ g : 1 \amalg A \to 1 \amalg B)$, then $e_A \leqslant u_1$ and the subobject u'_1, which is complementary to u_1, is such that $u'_1 \leqslant i_A$.

It follows that the fibred product of $(i_B : B \to 1 \coprod B, g : 1 \coprod A \to 1 \coprod B)$ is of the form $(f : U \to B, i_A u : U \to 1 \coprod A)$. The morphism $(u,f) : A \to B$ then satisfies $E(u,f) = g$. The functor E is thus full. It is essentially surjective because, for any object (A,a) of $\mathbb{C}_*$, the morphism $a : 1 \to A$ is a regular monomorphism and forms with its complement $a' : A' \to A$, a sum (a,a') of 1 and A', thus $(A,a) \simeq E(A')$. The functor E is thus an equivalence of categories. The category $\mathbb{A} = \mathbb{C}\mathrm{ont}_{\aleph_0} [\mathbb{C}, \mathbb{E}\mathrm{ns}]$ is locally simple (Theorem 2.3.0.) and the category $\mathbb{C}^{op}$ is equivalent to the category of finitely presentable objects in $\mathbb{A}$. By Theorem 3.2.0, the category $\mathbb{A}^*$ is quasi locally simple and the category of finitely presentable objects of $\mathbb{A}^*$ is equivalent to $(\mathbb{C}^{op})^* \simeq (\mathbb{C}_*)^{op} \sim (\mathbb{P}\mathrm{art}(\mathbb{C}))^{op}$. It follows that the category $\mathbb{C}\mathrm{ont}_{\aleph_0} [\mathbb{P}\mathrm{art}(\mathbb{C}), \mathbb{E}\mathrm{ns}] \sim \mathbb{A}^*$ is quasi locally simple. ∎

3.3.1. Examples. The category $\mathbb{F}\mathrm{oncfin}$ which has, as its objects, finite sets, and partially defined functions as morphisms, is the category $\mathbb{P}\mathrm{art}(\mathbb{E}\mathrm{nsfin})$ of partial morphisms of the category, $\mathbb{E}\mathrm{nsfin}$, of finite sets. The category $\mathbb{C}\mathrm{ont}_{\aleph_0} [\mathbb{F}\mathrm{oncfin}, \mathbb{E}\mathrm{ns}]$ is equivalent to the category of non-unitary Boolean algebras. One can replace $\mathbb{E}\mathrm{nsfin}$ by any small Boolean topos [23] and, in particular, by any small atomic topos [3].

3.4. LOCALLY BOOLEAN SHEAVES OF SIMPLE OBJECTS.

A locally Boolean space is a topological space in which each point has a neighbourhood which is a Boolean space. Also called generalised Boolean spaces [18], they can be defined as being those locally compact separated topological spaces which have a topological base formed of open-closed subsets. The category $\mathbb{E}\mathrm{spLocBool}$ of locally Boolean spaces has, as its objects, the locally Boolean spaces, and, as its morphisms from X to Y, the proper continuous partially defined functions from X to Y i.e. the continuous mappings $f : D \to Y$ from an open subset D of X to Y such that the inverse image of each compact subset of Y is a compact subset of X.

3.4.0. Proposition. The category, $\mathbb{E}\mathrm{spLocBool}$, of locally Boolean spaces is equivalent to the dual of the category, $\mathbb{B}\mathrm{oolNu}$, of non-unitary Boolean algebras.

Proof : This is a generalisation of the duality theorem of Stone for Boolean spaces [18]. Consider, on the one hand, the functor $B : \mathbb{E}spLocBool^{op} \to \mathbb{B}oolNu$, which associates to a locally Boolean space X, the non-unitary Boolean algebra, $B(X)$, of open-compact subsets of X, and, to a morphism $f : X \to Y$, the homomorphism $B(f) : B(Y) \to B(X)$ defined by $B(f)(u) = f^{-1}(u)$, and, on the other hand, the functor $S : \mathbb{B}oolNu \to \mathbb{E}spLocBool^{op}$, which associates to a non-unitary Boolean algebra, B, its Stone space, $S(B)$, that is the set of prime filters of B together with the topology generated by the basic open sets $D(a) = \{\Phi \in S(B) : a \in \Phi\}$ [18], and which associates, to a homomorphism $h : B \to C$, the function $S(h) : S(C) \to S(B)$ defined on the open set $D(h) = \{\Phi \in S(C) : \Phi \cap h(B) \neq \emptyset\}$ by $S(h)(\Phi) = h^{-1}(\Phi)$. One shows that these two functors are equivalences of categories quasi-inverse to each other [c.f. 18]. ■

Let $\mathbb{A}$ be a locally finitely presentable category. If B is a non-unitary Boolean algebra, we will also denote by B, the category associated to the ordered set B, with the structure of a site [21. 11.1.15] in which the Grothendieck topology T is generated by the pretopology formed of finite families $(a_1 \to a, \ldots, a_n \to a)$ such that $\bigvee_{i=1}^{n} a_i = a$, including amongst them the empty family covering 0. A presheaf $F : B^{op} \to \mathbb{A}$ is a sheaf if, and only if, $F0 \simeq 1$ and, for each pair (a,b) of elements of B such that $a \wedge b = 0$, the pair $(F(a \vee b) \to Fa, F(a \vee b) \to Fb)$ is a product in $\mathbb{A}$. It suffices, in fact, to note that the proof of [15, proposition 14.0], which is stated for Boolean algebras, uses, in fact, only relative complements and consequently is valid for non-unitary Boolean algebras. The category $\mathfrak{S}hvLocBool\,\mathbb{A}$, of locally Boolean sheaves of objects of $\mathbb{A}$ has, as its objects, pairs (B,F) where B is a non-unitary Boolean algebra and $F : B^{op} \to \mathbb{A}$ is a sheaf on B with values in $\mathbb{A}$, and, as its morphisms from (B,F) to (C,G), pairs (f,α) formed of a homomorphism of non-unitary Boolean algebras $f:B \to C$ and a natural transformation $\alpha : F \to Gf^{op}$.

The category $\mathfrak{S}hvEspLocBool\,\mathbb{A}$ of sheaves on locally Boolean spaces with values in $\mathbb{A}$ has, as its objects, pairs (X,F) where X is a locally Boolean space and F is a sheaf on X with values in $\mathbb{A}$, and, as its morphisms from (X,F) to (Y,G), pairs (f,α) where $f : Y \to X$ is a morphism of locally Boolean spaces and $\alpha : F \to Gf^*$ is a natural transformation, where $f^* : \Omega^{op}(X) \to \Omega^{op}(Y)$ is defined by $f^*(u) = f^{-1}(u)$.

3.4.1. Proposition. The categories $\mathfrak{S}$hvLocBool $\mathbb{A}$ and $\mathfrak{S}$hvEspLocBool $\mathbb{A}$ are equivalent.

Proof : It is immediate that the functor B : $\mathbb{E}$spLocBool^{op} → $\mathbb{B}$oolNu induces a functor, again denoted by B : $\mathfrak{S}$hvEspLocBool $\mathbb{A}$ → $\mathfrak{S}$hvLocBool $\mathbb{A}$, which is an equivalence of categories because the open-compact subsets of a locally Boolean space form a base for the topology and it suffices to define sheaves on the open sets of a base. The functor S : $\mathbb{B}$oolNu → $\mathbb{E}$spLocBool^{op} similarly induces a functor, again denoted S : $\mathfrak{S}$hvLocBool $\mathbb{A}$ → $\mathfrak{S}$hvEspLocBool $\mathbb{A}$, which is an equivalence of categories, quasi-inverse to that above. ∎

If $\mathbb{A}$ is a locally simple category (2.1.0), the category $\mathfrak{S}$hvLocBoolSim $\mathbb{A}$ of locally Boolean sheaves of simple objects of $\mathbb{A}$ is the full subcategory of $\mathfrak{S}$hvLocBool $\mathbb{A}$, having as its objects, the pairs (B,F) such that the stalks of F are simple objects.

3.4.2. Theorem. If $\mathbb{A}$ is a locally simple category, the category $\mathfrak{S}$hvLocBoolSim $\mathbb{A}$ of locally Boolean sheaves of simple objects of $\mathbb{A}$, is quasi locally simple.

Proof : Let $\mathbb{FL}$ denote the category $\mathfrak{S}$hvLocBoolSim $\mathbb{A}$

a) Let us start by studying the limits and colimits in $\mathbb{FL}$. On the boolean algebra $\{0\}$, there is a single sheaf F_0 and the pair $(\{0\}, F_0)$ is the null object of $\mathbb{FL}$. A morphism (f,α) of $\mathbb{FL}$ is null if, and only if, the homomorphism f is null. The kernel of a morphism $(f,\alpha) : (B,F) \to (C,G)$ is the pair (N,F_N) where N is the kernel of f and F_N is the restriction of F to N. Any normal subobject (Def. 3.5.1) of (B,F) is thus isomorphic to a subobject of the form $(N,F_N) \to (B,F)$, where N is an ideal of B. Let $(B_i,F_i)_{i\in\mathbb{I}}$ be a normal monomorphic diagram in $\mathbb{FL}$. (c.f. notation 3.6.). One may suppose that, for each morphism $u : i \to j$ in $\mathbb{I}$, the homomorphism $B_u : B_i \to B_j$ is the inclusion of an ideal B_i into B_j and that $F_u : F_i \to F_j B_u^{op}$ is the identity. Let $B = \varinjlim_{i\in\mathbb{I}} B_i$ in $\mathbb{B}$oolNu. One may suppose that the B_i are ideals of B and that B is a filtered union of the B_i. The functors $F_i : B_i^{op} \to \mathbb{A}$ define a functor $F : B^{op} \to \mathbb{A}$, which is a sheaf on B with values in $\mathbb{A}$, whose stalks are those of the F_i, thus are simple. The pair (B,F) is an object of $\mathbb{FL}$ and is the filtered colimit of the diagram $(B_i,F_i)_{i\in\mathbb{I}}$.

b) Let us denote by $\mathbb{F}$ the category $\mathbb{S}hvBoolSim\,\mathbb{A}$ of Boolean sheaves of simple objects of $\mathbb{A}$ (2.7). Any object (B,F) of $\mathbb{FL}$ is a normal monomorphic filtered colimit (c.f. notation 3.6.) of a diagram whose objects are in $\mathbb{F}$, since $(B,F) = \varinjlim_{b\in B}((b),F_{(b)})$, where (b) is the ideal of B generated by $b \in B$. Moreover, for any object (C,G) of $\mathbb{F}$ and any normal monomorphic filtered colimit $((f_i,\alpha_i) : (B_i,F_i) \to (B,F))_{i\in\mathbb{I}}$, any morphism $(f,\alpha) : (C,G) \to (B,F)$ factorises through one of the morphisms (f_i,α_i), since there is an $i \in \mathbb{I}$ such that $f(1) \in B_i$; consequently, the functor, $Hom_{\mathbb{FL}}((C,G),-)$ preserves normal monomorphic filtered colimits. Conversely, if an object (B,F) of $\mathbb{FL}$ is such that the functor, $Hom_{\mathbb{FL}}((B,F),-)$ preserves normal monomorphic filtered colimits, then the morphism $1_{(B,F)} : (B,F) \to (B,F) = \varinjlim_{b\in B}((b),F_{(b)})$ must factorise through a canonical injection $((b),F_{(b)}) \to (B,F)$ and consequently $(B,F) \simeq ((b),F_{(b)})$ is an object of $\mathbb{F}$. It follows that the objects of $\mathbb{F}$ are precisely the objects (B,F) of $\mathbb{FL}$ such that the functor, $Hom_{\mathbb{FL}}((B,F),-)$, preserves normal monomorphic filtered colimits. Denoting by $\mathbb{FP}$ the full subcategory of $\mathbb{FL}$ whose objects are those of $\mathbb{F}$ and by $k : \mathbb{FP} \to \mathbb{FL}$ the inclusion functor, the above properties imply that the fully faithful functor k is dense by normal monomorphic filtered colimits [8, Definition 2.0.].

c) Let us next study finite products in $\mathbb{FL}$. The product of two objects (B,F) and (C,G) is the pair $((p_1,\alpha_1) : (B\times C, F \otimes G) \to (B,F), (p_2,\alpha_2) : (B\times C, F \otimes G) \to (C,G))$ where $(p_1 : B\times C \to B,\ p_2 : B\times C \to C)$ is a product in $\mathbb{B}oolNu$, and $(\alpha_{1(b,c)} : F \otimes G(b,c) \to Fb\ , \alpha_{2(b,c)} : F\otimes G(b,c) \to Gc)$ is a product in $\mathbb{A}$. Any sheaf H on $B\times C$ is necessarily of the form $H = F \otimes G$, where F is a sheaf on B defined by $F(b) = H(b,0)$ and G is a sheaf on C defined by $G(c) = H(0,c)$, thus, any object of the form $(B\times C,H)$ is necessarily of the form $(B,F)\times(C,G)$. Moreover, any morphism of $\mathbb{F}$ of the form $(h,\gamma) : (B,F)\times(C,G) \to (D,H)$ is necessarily of the form $(h,\gamma) = (h_1,\gamma_1)\times(h_2,\gamma_2)$ with $(h_1,\gamma_1) : (B,F) \to (D_1,H_1)$ and $(h_2,\gamma_2) : (C,G) \to (D_2,H_2)$ where D_1 is the ideal of D generated by $h(1,0)$, D_2 is the ideal generated by $h(0,1)$, and H_1 and H_2 are restrictions of H.

d) Let us show that $\mathbb{F}$ is a reflexive subcategory of $\mathbb{FP}$. The initial object of $\mathbb{F}$ is the pair $(\Delta(Z),\tilde{Z})$, the structural sheaf of the initial object Z of $\mathbb{A}$. Let us show that, for each object (B,F) of $\mathbb{FP}$, the morphism $\eta = (1,0) : (B,F) \to (B,F)\times(\Delta(Z),\tilde{Z})$ is a universal morphism from (B,F) to the subcategory $\mathbb{F}$. Let $(g,\beta) : (B,F) \to (D,H)$

be a morphism of $\mathbb{FP}$. Denote by D_1 (resp. D_2) the ideal of D generated by $g(1)$ (resp. the complementary element of $g(1)$) and $g_1 : B \to D_1$ the Boolean homomorphism induced by g. Then, one has $D = D_1 \times D_2$, $g = (g_1, 0) : B \to D = D_1 \times D_2$, $(D,H) = (D_1,H_1) \times (D_2,H_2)$, and $(g,\beta) = ((g_1,\beta_1),(0,\beta_o)) : (B,F) \to (D_1,H_1) \times (D_2,H_2)$. On writing $(f_o,\alpha_o) : (\Delta(Z),\tilde{Z}) \to (D_2,H_2)$ for the uniquely defined morphism of $\mathbb{F}$, one obtains a morphism $(g_1,\beta_1) \times (f_o,\alpha_o) : (B,F) \times (\Delta(Z),\tilde{Z}) \to (D_1,H_1) \times (D_2,H_2)$ of $\mathbb{F}$ whose composite with the morphism η is (g,β). Any other morphism $(f,\alpha) : (B,F) \times (\Delta(Z),\tilde{Z}) \to (D,H)$ satisfying $(f,\alpha)\eta = (g,\beta)$ is such that $f(1,0) = g(1)$ and $f(0,1) = g(1)'$; thus (f,α) is of the form $(f_1,\alpha_1) \times (f_2,\alpha_2)$ with $(f_1,\alpha_1) : (B,F) \to (D_1,H_1)$ necessarily equal to (g_1,β_1) and with $(f_2,\alpha_2) : (\Delta(Z),\tilde{Z}) \to (D_2,H_2)$ necessarily equal to (f_o,α_o). This completes the proof that η is an universal morphism from (B,F) towards $\mathbb{F}$.

e) The reflection of $\mathbb{FP}$ in $\mathbb{F}$ generates the comonad $\mathbb{G} = (G,\varepsilon,\nu)$ on $\mathbb{F}$ where $G(-) = (-) \times (\Delta(Z),\tilde{Z})$, ε is the canonical projection, and ν is the product with the diagonal of $(\Delta(Z),\tilde{Z})$. The category of $\mathbb{G}$-coalgebras is equivalent to the category $\mathbb{F}^*$ of augmented objects of $\mathbb{F}$ (3.2.). The category $\mathbb{FP}$, as it has the same objects as $\mathbb{F}$, is thus isomorphic to the Kleisli category of $\mathbb{G}$ [33, p. 143] and consequently, there is a comparison functor $J : \mathbb{FP} \to \mathbb{F}^*$. The category $\mathbb{F}$ is locally simple (Theorem 2.7.0), thus the category $\mathbb{F}^*$ is quasi locally simple (Theorem 3.2.0) and the functor, J, induces an equivalence between the category $\mathbb{FP}$ and the category of objects of finite type of $\mathbb{F}^*$ (Theorem 3.6.7.). By Proposition 3.6.4., the functor J is dense by normal monomorphic filtered colimits.

f) The left Kan extension $E : \mathbb{FL} \to \mathbb{F}^*$ of J along k, and the left Kan extension $K : \mathbb{F}^* \to \mathbb{FL}$ of k along J, exist and preserve normal monomorphic filtered colimits [8, Theorem 2.1.]. The composites, $KE : \mathbb{FL} \to \mathbb{FL}$ and $EK : \mathbb{F}^* \to \mathbb{F}^*$ are left Kan extensions, respectively, of k along k and, of J along J [8, Theorem 2.1.], thus they are equivalent to identity functors. The categories $\mathbb{FL}$ and $\mathbb{F}^*$ are thus equivalent. and it follows that the category $\mathbb{FL}$ is quasi locally simple. ∎

Let $\mathbb{T}$ be an algebraic theory in the sense of Lawvere [30] having two distinguished constants, denoted 0 and 1. The category $\mathcal{S}hvLocBoolTsep\mathcal{A}lg(\mathbb{T})$ of <u>totally separated locally Boolean sheaves of</u>

$\mathbb{T}$-algebras is the subcategory of $\$hvLocBool\mathbb{A}lg(\mathbb{T}) \sim \$hvEspLocBool\mathbb{A}lg(\mathbb{T})$, whose objects are the pairs (X,F) with F totally separated (2.12.0.) and whose morphisms are the pairs (f,α) such that the morphisms on stalks, α_y, are monomorphisms.

3.4.3. Theorem. If $\mathbb{T}$ is an algebraic theory having two distinguished constants, the category, $\$hvLocBoolTsep\mathbb{A}lg(\mathbb{T})$, of totally separated locally Boolean sheaves of $\mathbb{T}$-algebras, is quasi locally simple, and its simple objects can be identified with the non-null $\mathbb{T}$-algebras.

Proof : The algebraic theory $\mathbb{C}\mathbb{T}$, obtained by adjoining to $\mathbb{T}$ a comparison operation, is locally simple, and the category $\mathbb{A}lg(\mathbb{C}\mathbb{T})$ is locally simple (Theorem 2.12.3.). By Theorem 3.4.2., the category $\$hvLocBoolSim\mathbb{A}lg(\mathbb{C}\mathbb{T})$ is quasi locally simple. However, this category is equivalent to the category $\$hvLocBoolTsep\mathbb{A}lg(\mathbb{T})$, since the proof of theorem 2.12.3. is valid for locally Boolean spaces. ∎

3.5. NORMAL SUBOBJECTS.

We will be working in a quasi locally simple category $\mathbb{A}$.

3.5.0. Proposition. For a morphism $m : M \to A$, the following assertions are equivalent :

(1) m is a monomorphism.

(2) The kernel of m is null.

(3) For any morphism $e : Z \to M$, one has that $me = 0 \Rightarrow e = 0$.

Proof : As the implications (1) => (2) => (3) are immediate, let us prove (3) => (1). Let X be a finitely presentable object and, $f,g : X \rightrightarrows M$, a pair of morphisms satisfying $mf = mg$. The coequaliser $k : M \to K$ of (f,g) is a finitely generated regular epimorphism, hence is the cokernel of a morphism $e : Z \to M$ (3.1.1.). The morphism k factorises m, hence $me = 0$ and consequently $e = 0$. Thus k is an isomorphism and $f = g$. Since the finitely presentable objects form a generating set for the category, it follows that m is a monomorphism. ∎

3.5.1. Definition. A monomorphism (resp. a subobject) is normal if it is the kernel of some morphism or, what amounts to the same, if it is the kernel of its cokernel.

In categories of k-algebras, the normal subobjects of A are the ideals of A and, in categories of locally Boolean sheaves of simple objects, the normal subobjects of (B,F) are the pairs (I,F_I) where I is an ideal of B and F_I is the restriction of F to I.

3.5.2. Proposition. The class of normal monomorphisms is stable under filtered unions and intersections.

Proof : Let $(n_i : N_i \to A)_{i\in I}$ be a family of normal monomorphisms with target A. For each $i \in I$, denote by $q_i : A \to Q_i$ the cokernel of n_i. The kernel of the morphism $(q_i)_{i\in I} : A \to \prod_{i\in I} Q_i$ is the intersection of the family $(n_i)_{i\in I}$. If the family $(n_i)_{i\in I}$ is filtered, the family of quotient objects, $(q_i)_{i\in I}$ is cofiltered and its cointersection is a filtered colimit, whose kernel $n : N \to A$ is the filtered colimit of the family $(n_i)_{i\in I}$, hence is the filtered union of the family $(n_i)_{i\in I}$. ∎

We will denote by N(A) the set of normal subobjects of an object A, ordered by the order induced from that of $S(A)$, and by $A/n : A \to A/n$ the cokernel of a normal monomorphism $n : N \to A$.

3.5.3. Proposition. There is a decreasing one-to-one correspondence between the set $N(A)$ of normal subobjects of A and the set, $R(A)$ of regular quotient objects of A, defined by associating to a normal subobject of A, its cokernel, and to a regular quotient object of A, its kernel.

Proof : The mapping $Q : N(A) \to R(A)$ defined by $Q(n) = A/n$ is decraasing, similarly is the mapping $K : R(A) \to N(A)$ defined by $K(q) = \ker(q)$. Any normal subobject being the kernel of its cokernel, one has $KQ = 1_{N(A)}$. Let $p : A \to P$ be a regular quotient object of A. It is a cofiltered cointersection, i.e. filtered colimit, of finitely generated regular quotient objects of A, [16], say $p = \bigcap_{i\in I} p_i$. Each morphism p_i is the cokernel of a morphism $e_i : Z \to A$, (3.1.1.), thus p is the cokernel of the morphism $e = \langle e_i \rangle_{i\in I} : \coprod_{i\in I} Z \to A$. Then p is cokernel of its kernel, which proves the relation : $QK = 1_{R(A)}$. ∎

3.5.4. Proposition. The complemented normal monomorphisms are precisely the morphisms of the form $(1,0) : X \to X\times Y$ and the corresponding complementary normal monomorphism is $(0,1) : Y \to X\times Y$. The complemented regular epimorphisms are precisely the morphisms of the form $1\times 0 : X\times Y \to X$ and the corresponding complement is $0\times 1 : X\times Y \to Y$.

Proof : Let us firstly show the second part of the statement. It is immediate that the product of two objects X and Y is a pair $(1\times 0 : X\times Y \to X,\ 0\times 1 : X\times Y \to Y)$ of two complemented regular epimorphisms. Let $(q : A \to Q,\ q' : A \to Q')$ be a pair of complemented regular epimorphisms. Denote by $(p : Q\times Q' \to Q,\ \ p' : Q\times Q' \to Q')$ the product of Q and Q' and write $k : A \to Q\times Q'$ for the morphism (q,q'). Let $(r : Q \to R,\ \ m : Q\times Q' \to R)$ be the amalgamated sum of (q,k), $(n : R \to N,\ s : Q \to N)$ the amalgamaged sum of (m,p) and $(n' : R \to N',\ s' : Q' \to N')$, the amalgamated sum of (m,p'). Then (nr,s) is the amalgamated sum of (q,q), thus $nr \simeq 1_Q$ and $s \simeq 1_Q$; similarly $(n'r,s')$ is the amalgamated sum of (q,q') thus $N' \simeq 0$. Since (p,p') is a counion of regular quotient objects, the pair $(n;R \to Q,\ \ n' : R \to 0)$ is a counion of regular quotient objects. Consequently, $n \simeq 1_{Q'}r \simeq 1_{Q'}m \simeq p$. It follows that $(1_Q,p)$ is the amalgamated sum of (q,k). One shows analoguously that $(1_{Q'},p')$ is the amalgamated sum of (q',k). Let $(t : Q\times Q' \to T,\ \ u : Q\times Q' \to T)$ be the amalgamated sum of (k,k), $(v : Q \to V,\ \ w : T \to V)$ the amalgamated sum of (p,t) and $(v' : Q' \to V',\ w' : T \to V')$ the amalgamated sum of (p',t). Then (v,wu) is the amalgamated sum of (q,k) ; so $v = 1_Q$ and $wu = p$; similarly, $(v',w'u)$ is the amalgamated sum of (q',k), thus $v' = 1_{Q'}$ and $w'u = p'$. The morphism $(w,w') : T \to Q\times Q'$ satisfies the relation : $(w,w')u = 1_{Q\times Q'}$ since $p(w,w')u = wu = p$ and $p'(w,w')u = w'u = p'$, and it also satisfies the relation $u(w,w') = 1_T$, since $wu(w,w') = p(w,w') = w$ and $w'u(w,w') = p'(w,w') = w'$ and the pair (w,w') is monomorphic, being a counion of regular epimorphisms, in fact the direct image along $tk = uk$ of the counion (q,q'). It follows that u is an isomorphism. Similarly, one shows that t is an isomorphism. It follows that k is epimorphic, hence an isomorphism, and thus that (q,q') is a product.

For the first part, consider a pair $(n : N \to A,\ \ n' : N' \to A)$ of complementary normal monomorphisms. The regular quotient objects $A/n : A \to A/n$, $A/n' : A \to A/n'$ are complementary (Proposition 3.5.3.) and thus form a product. The relations $n = \ker(A/n)$, and $n' = \mathrm{Ker}(A/n')$ imply that $N \simeq A/n'$, $N' \simeq A/n$, and the morphisms $(A/n)n' : N' \to A/n$ and $(A/n')n : N \to A/n'$ are isomorphisms. It follows that $n = (1,0) : N \to N\times N' = A$ and $n' = (0,1) : N' \to N\times N' = A$. ■

3.5.5. <u>Proposition</u>. <u>Any complemented normal monomorphism is split and has a unique retraction, and this retraction is a direct factor. Any direct factor is the retraction of one, and only one, normal monomorphism, and that normal monomorphism is complemented</u>.

<u>Proof</u> : Any complemented normal monomorphism, being of the form $n = (1,0) : X \to X \times Y$, is split by the direct factor $p = 1 \times 0 : X \times Y \to X$. Let $r : X \times Y \to X$ be another retraction of n. The morphism $q = 0 \times 1 : X \times Y \to Y$ is the cokernel of n, thus its direct image by r is the cokernel of $rn = 1_X$, that is to say that it is null. The direct image of p by r is thus 1_X and, consequently, r factorises through p in a morphism $s : X \to X$ satisfying $s = spn = rn = 1_X$, thus $r = p$. Any direct factor, being of the form $p = 1 \times 0 : X \times Y \to X$ is a retraction of the complemented normal monomorphism $n = (1,0) : X \to X \times Y$. Let $m : X \to X \times Y$ be another normal monomorphism satisfying $pm = 1_X$. It is of the form $m = (1,t)$ with $t : X \to Y$. Let $k : Y \to K$ be the cokernel of t and let $q = 0 \times 1 : X \times Y \to Y$. Let us show that $kq : X \times Y \to K$ is the cokernel of m. Any morphism $u : X \times Y \to U$ which satisfies $um = 0$ is such that the direct image of p by u is null, thus the direct image of q by u is 1_U i.e. u factorises through q in a morphism $v : Y \to U$ satisfying $vt = vqm = um = 0$; the morphism v thus factorises through k in a morphism $w : K \to U$ satisfying $wkq = vq = u$. The kernel of kq is of the form $X \times Y_1$ where Y_1 is the kernel of k. However it must be isomorphic to X. Thus $X \times Y_1 \simeq X$ and $Y_1 \simeq 0$ and it follows that k is an isomorphism, thus $t = 0$ and $m = (1,0)$. ∎

3.5.6. <u>Definition</u>. The intersection of the normal monomorphisms (resp. subobjects) greater or equal to a given monomorphism (resp. subobject) m, is the <u>normal monomorphism (resp. subobject) generated</u> by m.

It is the kernel of the cokernel of m because, if q is the cokernel of m, and n is the kernel of q, then $m \leqslant n$ and, for any normal monomorphism r such that $m \leqslant r$, one has $A/r \leqslant A/m = q = A/n$, so $n \leqslant r$.

3.5.7. <u>Definition</u>. <u>A normal monomorphism (resp. subobject) is of finite type</u> if it is generated by a monomorphism whose source is a finitely generated object.

It is precisely a normal monomorphism whose cokernel is finitely generated. It is thus complemented (Proposition 3.5.3. and 3.5.4).

Denote by $N_{tf}(A)$ the set of normal subobjects of A, which are of finite type.

3.5.8. Proposition. The set $N(A)$ of normal subobjects of A is a compactly generated complete Heyting algebra in which the compact elements are the normal subobjects of finite type. These are complemented and make up a non-unitary Boolean algebra $N_{ff}(A)$, which is a sublattice of $N(A)$.

Proof : The set $S(A)$ of subobjects of A is a complete lattice. $N(A)$ is a subset of $S(A)$ closed under intersections, thus is a complete lattice. The normal subobjects of finite type are compact [18] in $N(A)$, since, if n is one of them and $\cup_{i\in I} n_i$ is a filtered union of normal subobjects satisfying $n \leqslant \cup_{i\in I} n_i$, then n is generated by a finitely generated subobject m (definition 3.5.6.) and the relation $m \leqslant \cup_{i\in I} n_i$ implies there exist $i \in I$, $m \leqslant n_i$, and it follows that there is an $i \in I$ with $n \leqslant n_i$. Any object is a monomorphic filtered colimit of finitely generated objects [16], thus any subobject of A is the filtered union of finitely generated subobjects. Consequently, any normal subobject is the filtered union of normal subobjects of finite type. It follows that the lattice $N(A)$ is compactly generated, and, that the compact elements are precisely the normal subobjects of finite type and thus they are complemented. To prove that $N(A)$ is a complete Heyting algebra, there remains to prove the distributivity property, and it suffices to show this for the compact elements, but this can be done whilst we are proving that the complemented normal subobjects form a Boolean algebra which is a sublattice of $N(A)$.

Consider a complemented normal monomorphism $n = (1,0) : X \to X\times Y$ (Proposition 3.5.4.) and a normal monomorphism $u : U \to X\times Y$. The morphism u is of the form $u = (u_1,u_2)$ with $u_1 : U \to X$ and $u_2 : U \to Y$. Let $k : X\times Y \to K$ be the cokernel of u, $k_1 : X \to K_1$, the cokernel of u_1, and $k_2 : Y \to K_2$, the cokernel of u_2. The direct image of the product $(p : X\times Y \to X,\ q : X\times Y \to Y)$ by the epimorphism k, is a disjoint counion of regular epimorphisms, and thus, is a product (Proposition 3.5.4.), so $K = K_1\times K_2$ and $k = k_1\times k_2$. If $v : V \to X$ is the kernel of k_1 and $w : W \to Y$ is the kernel of k_2, then the kernel of $k = k_1\times k_2$ is equal to $v\times w$, so the monomorphism u is of the form

$v \times w$. The intersection of the two monomorphisms n and u is then the monomorphism $(v,0) : V \to X \times Y$. Their union is the monomorphism $1_X \times w : X \times W \to X \times Y$, since the cokernel of $1_X \times w$ is the quotient object $k_2 q : X \times Y \to K_2$, which is the cointersection of the cokernel $k_1 \times k_2 : X \times Y \to K_1 \times K_2$ of u and the cokernel $q : X \times Y \to Y$ of n. Consequently, if u is a complemented normal monomorphism, then so are v and w, $n \cap u$ and $u \cup n$. This proves that the complemented normal subobjects are stable under finite unions and finite intersections. Moreover if u is a normal monomorphism with $n \cap u = 0$, then $v = 0$, and u can be factorised via the complementary $(0,1) : Y \to X \times Y$ of n. The complement of a complemented normal subobject is thus also its pseudocomplement [18]. It follows that the complemented normal subobjects of A form a Boolean algebra [18, Theorem 4, section 6], which is a sublattice of $N(A)$.

The normal subobjects of finite type of A are stable under finite unions (Definition 3.5.7. and Proposition 3.5.3.). Let us show that they are also stable under non-empty finite intersections. Let n be a normal subobject of finite type of A and m a complemented normal subobject of A. If $\bigcup_{i \in I} n_i$ is a filtered union of normal subobjects of A such that $n \cap m \leqslant \bigcup_{i \in I} n_i$, then $n = n \cap (m \cup m') = (n \cap m) \cup (n \cap m') \leqslant (\bigcup_{i \in I} n_i) \cup m' = \bigcup_{i \in I}(m' \cup n_i)$, hence there is an index $i \in I$ such that $n \leqslant m' \cup n_i$ and hence $n \cup m \leqslant (m' \cup n_i) \cap m = (m' \cap m) \cup (n_i \cap m) = n_i \cap m \leqslant n_i$. It follows that $n \cap m$ is a normal subobject of finite type of A. As a result, the normal subobjects of finite type of A form a sublattice $N_{tf}(A)$ of $N(A)$, which is bounded below and such that, for any pair of elements (n,m) of $N_{tf}(A)$ satisfying $m \leqslant n$, the element $m' \cap n \in N_{tf}(A)$ is the complement of m in the segment $[0,n]$. $N_{tf}(A)$ is thus a non-unitary Boolean algebra as well as being a sublattice of $N(A)$ [18]. ∎

3.5.9. <u>The functors $N : \mathbb{A} \to \mathbb{H}\text{eytcompNu}$ and $N_{tf} : \mathbb{A} \to \mathbb{B}\text{oolNu}$.</u>

<u>The category, $\mathbb{H}$eytCompNu, of non-unitary complete Heyting algebras</u> has, as its objects, the complete Heyting algebras and, as its morphisms, the mappings which preserve joins of pairs of elements and arbitrary meets. The functor $N : \mathbb{A} \to \mathbb{H}\text{eytCompNu}$ sends A to the complete Heyting algebra $N(A)$ of normal subobjects of A, and, a morphism $f : A \to B$, to the mapping $N(f) : N(A) \to N(B)$ which to n associates the kernel of the cokernel of fn. The mappings $N(f) : N(A) \to N(B)$ are morphisms in $\mathbb{H}$eytCompNu since they define, on passage to the corresponding cokernels, a mapping $R(A) \to R(B)$, which preserves arbitrary cointersections as well as counions of pairs of regular

quotient objects since these are couniversal. Moreover, if $n \in N(A)$ is of finite type, its cokernel is finitely generated, as is the direct image of A/n by f, and, consequently, $N(f)(n)$ is of finite type. Thus the functor N induces the functor $N_{tf} : \mathbb{A} \to \mathbb{B}\mathrm{oolNu}$.

3.5.10. Normal factorisation.

3.5.10.0. Proposition. The composite of two normal monomorphisms is a normal monomorphism.

Proof : Let $m : M \to N$ and $n : N \to A$ be two normal monomorphisms. If n and m are of finite type, they are of the form $(1,0) : X \to X\times Y$ (Propositions 3.5.4 and 3.5.8.), thus their composite nm is a normal monomorphism. If n is of finite type, then $m = \bigcup_{i\in I} m_i$ is a filtered union of normal monomorphisms of finite type, and the composite $nm = \bigcup_{i\in I} nm_i$ is a normal monomorphism as it is a filtered union of normal monomorphisms (Proposition 3.5.2.). In general, $n = \bigcup_{k\in K} n_k$ is a filtered union of normal monomorphisms of finite type, and, if one denotes by $r_k : N_k \to N$ the normal monomorphism which satisfies $nr_k = n_k$, and by $m_k : M_k \to N_k$ the inverse image of m by r_k, then $nm = nm \cap n = nm \cap (\bigcup_{k\in K} n_k) = \bigcup_{k\in K}(nm \cap n_k) = \bigcup_{k\in K} n_k m_k$ is a normal monomorphism as it is a filtered union of normal monomorphisms. ∎

3.5.10.1. Definition. A morphism $f : A \to B$ is a semi-epimorphism if its cokernel if null or, what amounts to the same, if for any morphism $g : B \to C$, one has : $gf = 0 \Rightarrow g = 0$.

3.5.10.2. Examples. In categories of K-algebras, the semi-epimorphisms are those algebra homomorphisms $f : A \to B$ such that the ideal of B generated by $f(A)$ is B. In particular, if A is a unitary K-algebra, the semi-epimorphisms $f : A \to B$ are precisely the unitary homomorphisms. In categories of locally Boolean sheaves of simple objects, the semi-epimorphisms are the morphisms $(f,\alpha) : (B,F) \to (C,G)$, where f is a semi-epimorphism of Boolean algebras. In categories of sheaves of simple objects on locally Boolean spaces, the semi-epimorphisms are precisely the pairs $(f,\alpha) : (X,F) \to (Y,G)$ where f is an everywhere defined function on Y.

3.5.10.3. Proposition. Semi-epimorphisms are exactly the morphisms which are "orthogonal" to normal monomorphisms. They are stable under composition. If gf is a semi-epimorphism, g is one too. If $f : A \to B$ and $g : A \to C$ are semi-epimorphisms, then (f,g) : A → B×C is a semi-epimorphism. If f : A → B, and h : D → C are semi-epimorphisms, then f×h : A×D → B×C is a semi-epimorphism.

Proof : Let f : A → B be a semi-epimorphism, n : N → C a normal monomorphism, g : A → N and h : B → C two morphisms satisfying hf = ng. The relation (C/n)hf = (C/n)ng = 0 implies (C/n)h = 0 and the existence of a unique morphism m : B → N satisfying h = nm and thus mf = g. Semi-epimorphisms are thus "orthogonal" to normal monomorphisms. Conversely, if f : A → B is a morphism "orthogonal" to normal monomorphisms, then the cokernel k : B → K of f, together with the morphism g = 0 : A → 0 and the normal monomorphism n = 0 : 0 → K satisfy kf = ng, thus there is a morphism m : B → 0 satisfying k = nm, i.e. k = 0. The morphism f is thus a semi-epimorphism. The stability of semi-epimorphisms under composition and right simplication is immediate. Let f : A → B and g : A → C be two semi-epimorphisms. Let (p : B×C → B, q : B×C → C) be the product of B and C, and h : B×C → D, a morphism satisfying h(f,g) = 0. Let (m : D → M, r : B → M) be the amalgamated sum of (h,p), and (n : D → N, s : C → B) the amalgamated sum of (h,q). Since (p,q) is a counion of regular quotient objects, the pair (m,n) is a counion of regular quotient objects, and thus is a monomorphic pair. The relation rf = rp(f,g) = mh(f,g) = 0 implies that r = 0, and the relation sg = sq(f,g) = nh(f,g) = 0 implies s = 0. It follows that mh = 0 = nh and hence h = 0. The morphism (f,g) : A → B×C is thus semi-epimorphic. Let k : D → C be a semi-epimorphism. If (u : A×D → A, v : A×D → D) is the product of A and D, the morphism f×k = (fu,kv) : A×D → B×C is semi-epimorphic by what has already been proved. ■

3.5.10.4. Proposition. Any morphism can be factorised in a natural and essentially unique way as the composite of a semi-epimorphism followed by a normal monomorphism. This factorisation is called the normal factorisation.

Proof : Consider a morphism f : A → B. Let q : B → S be its cokernel, n : N → B the kernel of q, and s : A → N the morphism satisfying ns = f. We will show that s is a semi-epimorphism. Let g : N → C be a morphism satisfying gs = 0. Let m : M → N be the kernel of g and h : A → M the morphism satisfying mh = s. Then nm : M → B is a normal monomorphism whose cokernel is q. It follows that m is an isomorphism, so g = 0. The naturality and the uniqueness up to isomorphism follow from Proposition 3.5.10.3. ■

3.5.10.5. Remark. In the normal factorisation $m = ns$ of a monomorphism m, n is the normal monomorphism generated by m. (Definition 3.5.6.).

3.6. OBJECTS OF FINITE TYPE.

We will work in a quasi locally simple category, $\mathbb{A}$.

A filtered colimit $(\alpha_i : A_i \to A)_{i \in \mathbb{I}}$ in which the canonical injections α_i are normal monomorphisms is called a normal monomorphic filtered colimit. The family of monomorphisms (resp. subobjects) $(\alpha_i : A_i \to A)_{i \in Ob(\mathbb{I})}$ is then a filtered union of normal monomorphisms (resp. subobjects), and any filtered union of normal monomorphisms (resp. subobjects) is of this form.

3.6.0. Definition. An object A is of finite type if the functor $Hom_{\mathbb{A}}(A,-) : \mathbb{A} \to \mathbb{E}ns$ preserves normal monomorphic filtered colimits or, equivalently, normal monomorphic filtered unions, that is to say, that, for any normal monomorphic filtered union $(n_i : N_i \to N)_{i \in I}$, any morphism $f : A \to N$ factorises through one of the morphisms n_i.

3.6.1. Examples. Finitely presentable and finitely generated objects are of finite type. In categories of K-algebras, the objects of finite type are exactly the unitary K-algebras. In categories of locally Boolean sheaves of simple objects, they are the Boolean sheaves (Part(b) of Theorem 3.4.2.).

An object B is said to be a semi-quotient of an object A, if there is a semi-epimorphism $f : A \to B$.

3.6.2. Proposition. The class of objects of finite type is stable under semi-quotients and coincides with the class of semi-quotients of finitely presentable objects.

Proof : Let A be an object of finite type and $f : A \to B$ a semi-epimorphism. If $(n_i : N_i \to N)_{i \in I}$ is a filtered union of normal monomorphisms and $g : B \to N$ is a morphism, there is an index, $i \in I$, and a morphism $g_i : A \to N_i$ satisfying $n_i g_i = gf$. From the orthogonality property of semi-epimorphisms, there is a morphism $h : B \to N_i$ satisfying $g = n_i h$. The object B is thus of finite type. Since finitely presentable objects are of finite type, any semi-quotient of a finitely presentable object is of finite type. Conversely, consider an object of finite type C. There is a filtered colimit $(\gamma_i : C_i \to C)_{i \in \mathbb{I}}$, where the objects C_i are finitely presentable. The normal factorisation $\gamma_i = n_i s_i$ of the morphisms γ_i determines a filtered union of normal monomorphisms $(n_i)_{i \in Obj(\mathbb{I})}$. The morphism, 1_C, factorises through one of the monomorphisms n_i which must therefore be an isomorphism. Hence the morphism γ_i must be a semi-epimorphism, so C is a semi-quotient of the finitely presentable object, C_i. ∎

3.6.3. <u>Proposition</u>. <u>A normal monomorphism is of finite type (Definition 3.5.7.) if, and only if, its source is of finite type</u>.

<u>Proof</u> : Since a normal monomorphism of finite type, $n : N \to A$ is generated by a monomorphism $m : M \to A$, where the source M of m is a finitely generated object, n appears in the normal factorisation $m = ns$, of m (Remark 3.5.10.5.), in company with a semi-epimorphism $s : M \to N$ which makes N a semi-quotient object of M, and thus an object of finite type. Conversely, a normal monomorphism $n : N \to A$, whose source is an object of finite type, is a filtered union of normal monomorphisms of finite type, say $n = \bigcup_{i \in I} n_i$. For each $i \in I$, let $m_i : N_i \to N$ be the morphism such that $nm_i = n_i$. The family $(m_i : N_i \to N)_{i \in I}$ is a normal monomorphic filtered union. The morphism 1_N must factorise through one of the morphisms m_i, which, as a result, must be an isomorphism, and so $n \simeq n_i$ is of finite type. ∎

3.6.4. <u>Proposition</u>. <u>The full subcategory of $\mathbb{A}$ having as objects, the objects of finite type, is dense by normal monomorphic filtered colimits</u> [8].

<u>Proof</u> : If k is the inclusion functor, the normal monomorphic filtered colimits are k-absolute [8], by the definition of objects of finite type. Moreover, any object A of $\mathbb{A}$ is a filtered colimit of finitely presentable objects : say $(A, (\alpha_i)) = \varinjlim_{i \in \mathbb{I}} A_i$. Each morphism $\alpha_i : A_i \to A$ can be factorised in the form $\alpha_i = n_i s_i$ with $s_i : A_i \to B_i$ a semi-epimorphism and $n_i : B_i \to A$ a normal monomorphism (Proposition 3.5.10.4.). Then B_i is of finite type (Proposition 3.6.2.) and $(n_i : B_i \to A)_{i \in \mathbb{I}}$ is a normal monomorphic filtered colimit. Thus any object of A is a normal monomorphic filtered colimit of objects of finite type. ∎

<u>The category $\mathbb{A}_{tf}$ of objects of finite type of $\mathbb{A}$</u> has, as its objects, the objects of finite type, and, as its morphisms, the semi-epimorphisms of $\mathbb{A}$.

3.6.5. <u>Proposition</u>. <u>The category $\mathbb{A}_{tf}$, of objects of finite type of $\mathbb{A}$, is a reflexive subcategory of $\mathbb{A}$, which is closed under connected colimits and whose initial object is the coclassifying object Z.</u>

<u>Proof</u> : a) As the regular epimorphism $Z \to 0$ is finitely generated (cf. 3.1.1.), its kernel $1_Z : Z \to Z$ is a normal monomorphism of finite type and it follows that the object, Z, is of finite type (Proposition 3.6.3.). If A is an object of finite type, the normal monomorphism $1_A : A \to A$ is of finite type, so its cokernel $A \to 0$ is a finitely generated regular epimorphism, and there is a unique morphism $e : Z \to A$ whose cokernel is null, that is, there is a unique semi-epimorphism $e : Z \to A$. We will denote it by e_A. The object, Z, is thus initial in $\mathbb{A}_{tf}$.

b) Let $(A_i)_{i\in\mathbb{I}}$ be a non-empty connected diagram in $\mathbb{A}_{tf}$ with colimit $(\alpha_i : A_i \to A)_{i\in\mathbb{I}}$ in $\mathbb{A}$. If $f : A \to B$ is a morphism satisfying $f\alpha_i = 0$ for some object i of $\mathbb{I}$, then, for any morphism $u : i' \to i$ of $\mathbb{I}$, one has $f\alpha_{i'} = f\alpha_i A_u = 0$, and, for any morphism $v : i \to i''$ of $\mathbb{I}$, one has $f\alpha_{i''}A_v = f\alpha_i = 0$ so $f\alpha_{i''} = 0$ because Av is a semi-epimorphism. Since the category $\mathbb{I}$ is connected, it follows that for any object j of $\mathbb{I}$, $f\alpha_j = 0$, and consequently $f = 0$. The canonical injections α_i are thus semi-epimorphisms and hence, the object A is of finite type. It follows that the cone $(\alpha_i : A_i \to A)_{i\in\mathbb{I}}$ is in the subcategory $\mathbb{A}_{tf}$, and, that it is a colimit in that category.

c) Consider first of all, an object, A, of finite type in $\mathbb{A}$. We will construct a universal morphism from A to the subcategory $\mathbb{A}_{tf}$. The morphism $(1,e_A) : Z \to Z\times A$ is a semi-epimorphism (Proposition 3.5.10.3), so the object $Z\times A$ is of finite type. We will show that the morphism $\eta = (0,1) : A \to Z\times A$ is universal from A to $\mathbb{A}_{tf}$. Let B be an object of finite type and $g : A \to B$. Let $g = nf$ be the normal factorisation of g, with $f : A \to X$, and $n : X \to B$. Since A is of finite type, X is of finite type (Proposition 3.6.2.). The morphism n is a normal monomorphism of finite type (Proposition 3.6.3.), so is of the form $n = (0,1) : X \to Y\times X$ (Proposition 3.5.4.). It follows that g is of the form $g = (0,f) : A \to Y\times X$ with X and Y of finite type and f a semi-epimorphism. The morphism $e_Y \times f : Z\times A \to Y\times X$ is a semi-epimorphism (Proposition 3.5.10.3.) and satisfies

$(e_Y\times f)\eta = (e_Y\times f)(0,1) = (0,f) = g$. Consider another semi-epimorphism $h : Z\times A \to Y\times X$ satisfying $h\eta = g$. Denote by $(p : Z\times A \to Z,\ q : Z\times A \to A)$ the product of Z and A, and by $(r : Y\times X \to Y,\ s : Y\times X \to X)$ the product of Y and X. Then p is the cokernel of η. The morphism rh satisfies $rh\eta = rg = r(0,f) = 0$, so it factorises through p in a morphism $Z \to Y$ which is necessarily semi-epimorphic, thus equal to e_Y. Let $(t : Y\times X \to T,\ u : A \to T)$ be the amalgamated sum of (h,q). Since r is the direct image of p by h, the pair (r,t) is the direct image by h of the pair (p,q). Thus (r,t) is a counion of regular quotient objects. It is a disjoint counion, since any morphism $v : Y\times X \to V$, which factorises through r and t, is such that vh factorises through p and q, thus $vh = 0$ and $v = 0$. It follows that r and t define complementary quotient objects and so $t \simeq s$ and $h = e_Y\times u : Z\times A \to Y\times X$. Then $(0,f) = (e_Y\times u)\eta = (e_Y\times u)(0,1) = (0,u)$ thus $f = u$, which proves that $h = e_Y\times f$. The morphism $\eta : A \to Z\times A$ is thus universal from A to $\mathbb{A}_{tf}$. One thus defines in a canonical way, a functor $T_o : \bar{\mathbb{A}}_{tf} \to \mathbb{A}_{tf}$ partial left adjoint to the inclusion functor, $\mathbb{A}_{tf} \to \mathbb{A}$, defined on the full subcategory $\bar{\mathbb{A}}_{tf}$ of $\mathbb{A}$ whose objects are the objects of finite type.

d) Any object A of $\mathbb{A}$ is a filtered colimit of objects of finite type, say $A = \varinjlim_{i\in\mathbb{I}} A_i$. Then the image under the functor $T_o : \bar{\mathbb{A}}_{tf} \to \mathbb{A}_{tf}$ of the diagram $(A_i)_{i\in\mathbb{I}}$, has a colimit TA in $\mathbb{A}$, which canonically comes with a morphism $\eta_A : A \to TA$, which is a universal morphism from A to $\mathbb{A}_{tf}$. The inclusion functor $\mathbb{A}_{tf} \to \mathbb{A}$, thus has a left adjoint, T. ∎

3.6.6. Theorem. If $\mathbb{A}$ is a quasi locally simple category, the category $\mathbb{A}_{tf}$, of objects of finite type of $\mathbb{A}$, is locally simple.

Proof : The category $\mathbb{A}_{tf}$ is cocomplete as it is a reflexive subcategory of the cocomplete category $\mathbb{A}$. As $\mathbb{A}_{tf}$ is closed, in $\mathbb{A}$, under filtered colimits, the objects which are images under the reflector $T : \mathbb{A} \to \mathbb{A}_{tf}$ of finitely presentable objects of $\mathbb{A}$, are finitely presentable in $\mathbb{A}_{tf}$, and form a set of proper generators for $\mathbb{A}_{tf}$. The category $\mathbb{A}_{tf}$ is thus locally finitely presentable. As the object Z is of finite type, it is a semi-quotient of a finitely presentable object, A. Therefore there is a semi-epimorphism $f : A \to Z$ which, with the semi-epimorphism $e_A : Z \to A$ satisfies the relation $fe_A = 1_Z$. Then f is a split epimorphism, and Z is finitely presentable in $\mathbb{A}$, as it is a split quotient of a finitely presentable object. The object

$TZ = Z\times Z = Z^2$ is thus finitely presentable in $\mathbb{A}_{tf}$. The object, Z^2, detects monomorphisms in $\mathbb{A}_{tf}$ since, if $f : A \to B$ is a morphism of $\mathbb{A}_{tf}$ such that $\mathrm{Hom}_{\mathbb{A}_{tf}}(Z^2,f)$ is injective, the mapping $\mathrm{Hom}_{\mathbb{A}}(Z,f) \simeq \mathrm{Hom}_{\mathbb{A}_{tf}}(TZ,f) = \mathrm{Hom}_{\mathbb{A}_{tf}}(Z^2,f)$ is injective, and it follows that the morphism f is a monomorphism (Proposition 3.5.0.). Finite products are codisjoint in $\mathbb{A}_{tf}$ since they are in $\mathbb{A}$ and the inclusion functor $\mathbb{A}_{tf} \to \mathbb{A}$ preserves and reflects products and amalgamated sums. The final object of $\mathbb{A}_{tf}$ is the null object of $\mathbb{A}$. It is a strict final object in $\mathbb{A}_{tf}$ since any semi-epimorphism $s : 0 \to A$ has a cokernel which is both 0 and 1_A, so $A = 0$. Let $(p : A\times B \to A,\ q : A\times B \to B)$ be the product of two objects A and B of $\mathbb{A}_{tf}$ and $f : A\times B \to C$ a morphism of $\mathbb{A}_{tf}$. Denote by $(g : A \to R,\ r : C \to R)$, the amalgamated sum of (p,f), $(h : B \to S,\ s : C \to S)$, the amalgamated sum of (q,f), and $(m : R \to M, n : S \to M)$ the amalgamated sum of (r,s). As the objects A and B are of finite type, the normal monomorphisms $(1,0) : A \to A\times B$ and $(0,1) : B \to A\times B$ are of finite type and their cokernels $q : A\times B \to B$ and $p : A\times B \to A$ are finitely generated regular epimorphisms. Their direct images s and r are finitely generated regular epimorphisms, and thus are complemented. They are direct factors (Proposition 3.5.4.) forming a counion $(r : C \to R,\ s : C \to S)$. The morphism mrf factorises via p since $mrf = mgp$, and via q since $mrf = nsf = nhq$, hence it is null. This implies that $ns = mr = 0$, so the counion (r,s) is codisjoint, and thus is a product. Finite products are thus couniversal in $\mathbb{A}_{tf}$. This completes the proof that $\mathbb{A}_{tf}$ is locally simple (Definition 2.1.0.). ∎

3.6.7. <u>Theorem</u>. *If $\mathbb{A}$ is a locally simple category, the functor $J : \mathbb{A} \to \mathbb{A}^*$ defined by $J(-) = (Z\times(-), p_{(-)})$ induces an equivalence between the category $\mathbb{A}$ and the category of objects of finite type of $\mathbb{A}^*$.*

<u>Proof</u> : Let $f : A \to B$ be a morphism of $\mathbb{A}$. We will show that $Jf = 1_Z\times f : (Z\times A,\ p_A) \to (Z\times B,\ p_B)$ is a semi-epimorphism in $\mathbb{A}^*$. Let $g : (Z\times B,\ p_B) \to (C,c)$ be a morphism such that $g(Jf) = 0$, i.e. $g(1_Z\times f) = i_C p_A$ where $i_C : Z \to C$ denotes the unique morphism. The direct image of the projection $q_A : Z\times A \to A$ by the morphism $g(1_Z\times f)$ is the morphism $t_C : C \to 1$, and hence, the direct image of the projection $q_B : Z\times B \to B$ by the morphism g is also t_C. The direct image of the projection $p_B : Z\times B \to Z$ by g is thus 1_C, i.e., $g = i_C p_B$, thus the

morphism $g : (Z\times B,\ p_B) \to (C,c)$ is null. As a result, the morphism Jf must be a semi-epimorphism. The functor J preserves finitely presentable objects ((c)) proof of Theorem 3.2.0). Since any object of $\mathbb{A}$ is a filtered colimit of finitely presentable objects and, as J preserves filtered colimits, any image under J of an object in $\mathbb{A}$ is a filtered colimit of finitely presentable objects, and is thus an object of finite type (Proposition 3.6.5.). The functor J thus induces a functor $J_{tf} : \mathbb{A} \to \mathbb{A}^*_{tf}$. This functor is faithful ((a) proof of Theorem 3.2.0.). We will show that it is full. Consider an object A of $\mathbb{A}$, an object (B,b) of $\mathbb{A}^*$ and a semi-epimorphism, $g : JA \to (B,b)$. Let $(q : B \to M,\ i_M : Z \to M)$ be the amalgamated sum of $(g : Z\times A \to B,\ p_A : Z\times A \to Z)$, and let $m : M \to Z$ be the morphism such that $mq = b$ and $mi_M = 1_Z$. The morphism $q : (B,b) \to (M,m)$ is null, since its composite with the morphism $g : (Z\times A,\ p_A) \to (B,b)$ is so. It follows that $q = i_M b$, so i_M is an isomorphism, as is m, and hence, $(b,1_Z)$ is the amalgamated sum of (g,p_A). As a result the amalgamated sum of $(g : Z\times A \to B,\ q_A : Z\times A \to A)$ is of the form $(q_B : B \to C,\ f : A \to C)$, where $(q_B : B \to C,\ b : B \to Z)$ is a product. The object (B,b) is thus of the form $J(C)$ and the morphism g of the form $g = 1_Z\times f = Jf$. This proves that the functor J_{tf} is full, since, if A' is an object of $\mathbb{A}$ such that $JA' = (B,b)$, then one has $A' \simeq C$ in our previously used notation. This proves also that the functor J_{tf} is essentially surjective, since, for any object of finite type (B,b) of $\mathbb{A}^*$, there is a semi-epimorphism $J(Z) = (Z^2,\Pi) \to (B,b)$ and hence the object (B,b) is of the form JC. It follows that J_{tf} is an equivalence of categories. ∎

3.7. MORPHISMS OF QUASI LOCALLY SIMPLE CATEGORIES.

3.7.0. Definition. A morphism of quasi locally simple categories is a functor which satisfies the following properties :

(i) its source and its target are quasi locally simple categories,

(ii) it reflects the null object,

(iii) it preserves filtered colimits,

(iv) it has a left adjoint which preserves finite products.

If $\mathbb{A}$ and $\mathbb{B}$ are two quasi locally simple categories, a functor $U : \mathbb{A} \to \mathbb{B}$ lifts subobjects (resp. normal subobjects) if for any object A of $\mathbb{A}$ and any monomorphism (resp. normal monomorphism)

$n : N \to UA$ of $\mathbb{B}$, there exists a monomorphism (resp. a normal monomorphism) $m : M \to A$ of $\mathbb{A}$ and an isomorphism $r : UM \to N$ satisfying $nr = Um$. It <u>lifts</u> them <u>uniquely</u> if, moreover, for any other monomorphism (resp. normal monomorphism) $m_1 : M_1 \to A$ of $\mathbb{A}$ such that there is an isomorphism $r_1 : UM_1 \to N$ satisfying $nr_1 = Um_1$, there is an isomorphism $s : M_1 \to M$ satisfying $ms = m_1$.

3.7.1. <u>Proposition</u>. <u>Let $\mathbb{A}$ and $\mathbb{B}$ be two quasi locally simple categories. For a functor $U : \mathbb{A} \to \mathbb{B}$ which preserves filtered colimits and has a left adjoint, the following assertions are equivalent:</u>

(1) <u>U is a morphism of quasi locally simple categories</u>,

(2) <u>U lifts normal subobjects and reflects the null object</u>,

(3) <u>U uniquely lifts normal subobjects</u>,

(4) <u>the adjunction morphisms of U are semi-epimorphisms</u>.

<u>Proof</u> : The functor U preserves kernels and hence normal monomorphisms. It preserves products, thus direct factors, that is to say, finitely generated regular epimorphisms, and by passing to filtered counions, it preserves all the regular epimorphisms. It preserves cokernels of normal monomorphisms since, if A/n is the cokernel of a normal monomorphism $n : N \to A$, then $U(A/n)$ is a regular epimorphism with kernel Un, and thus is the cokernel of Un.

(0) Firstly, let us show that U reflects the null object if and only if, it reflects semi-epimorphisms. Suppose that U reflects the null object. If $f : A \to B$ is a morphism of $\mathbb{A}$, whose image, Uf, is a semi-epimorphism, then the cokernel $q : B \to Q$ of f has, as its image under U, a regular epimorphism Uq which satisfies $(Uq)(Uf) = 0$, thus which is null, and so $UQ = 0$, $Q = 0$ and f is a semi-epimorphism. Conversely, if U reflects the semi-epimorphisms and A is an object of $\mathbb{A}$ such that $UA = 0$, then the normal monomorphism $0 \to A$ is semi-epimorphic since its image under U is, thus it is an isomorphism and hence $A = 0$.

<u>(1) => (2)</u>. Let us denote by (U,F,η,ε) the adjunction associated to U. The functor F preserves normal monomorphisms of finite type with their complements, since they are of the form $(1,0) : X \to X\times Y$ and F preserves finite products. The image under U of the morphism $\varepsilon_A : FUA \to A$ is a split epimorphism, thus is a semi-epimorphism. As the functor U reflects the null object, it reflects semi-epimorphisms by (0), thus ε_A is a semi-epimorphism. It follows that the homomorphism

of Boolean algebras, $N_{tf}(\varepsilon_A) : N_{tf}(FUA) \to N_{tf}(A)$ is unitary and that it preserves complements. For any normal monomorphism of finite type, $n : N \to UA$ with complement $n' : N' \to UA$, the normal monomorphisms, $m = N_{tf}(\varepsilon_A)(Fn)$ and $m' = N_{tf}(\varepsilon_A)(Fn')$ are thus complementary and satisfy $n \leqslant Um$ and $n' \leqslant Um'$, thus $n \simeq Um$ and $n' \simeq Um'$. The functor U thus lifts the normal subobjects of finite type. Since any normal subobject is a filtered union of normal subobjects of finite type, and, as the functor U preserves filtered colimits and thus filtered unions of normal subobjects, it follows that U lifts all the normal subobjects.

<u>(2) => (3)</u>. By (0), the functor U lifts semi-epimorphisms. Let $n : N \to A$ and $n_1 : N_1 \to A$ be two normal monomorphisms of $\mathbb{A}$ such that $Un \simeq Un_1$. Let $n_2 : N_2 \to A$ be the intersection of n and n_1 and $m : N_2 \to N$ the normal monomorphism satisfying $nm = n_2$. Then $Un_2 \simeq Un$, thus Um is an isomorphism, and it follows that m is a semi-epimorphism thus an isomorphism and one has $n \simeq n_2$. Similarly one has $n_1 \simeq n_2$ and hence $n \simeq n_1$. The functor, U, thus lifts normal subobjects uniquely.

<u>(3) => (4)</u>. The functor U reflects the null objects since, if A is an object of $\mathbb{A}$ such that $UA = 0$, the two normal monomorphisms $0 \to A$ and 1_A have the same image under U, thus are isomorphic and consequently $A = 0$. It follows from (0) that the morphisms $\varepsilon_A : FUA \to A$ are semi-epimorphisms since their images under U are split epimorphisms. For an object B of $\mathbb{B}$, the adjunction morphism η_B factorises normally in the form $\eta_B = ns$. The normal monomorphism n is of the form $n = Um$ where $m : M \to FB$ is a normal monomorphism. There exists a morphism $f : FB \to M$ satisfying $(Uf)\eta_B = s$. Then $U(mf)\eta_B = (Um)(Uf)\eta_B = (Um)s = \eta_B$, thus $mf = 1_{FB}$ and m is an isomorphism as is n. It follows that η_B is a semi-epimorphism.

<u>(4) => (1)</u>. If A is an object of $\mathbb{A}$ such that $UA = 0$, then $FUA = 0$, and the existence of a semi-epimorphism $\varepsilon_A : FUA \to A$ proves that $A \simeq 0$. The functor U thus reflects the null object. The functor F preserves the final object, since it is the null object. Let $(p : B\times C \to B,\ q : B\times C \to C)$ be the product of two objects B and C of $\mathbb{B}$. Since the morphism $p : B\times C \to B$ is the cokernel of the normal monomorphism $n = (0,1) : C \to B\times C$, the morphism $UFp : UF(B\times C) \to UB$ is the cokernel of the morphism $UFn : UFC \to UF(B\times C)$. Since $\eta_C : C \to UFC$ is a semi-epimorphism, the morphism UFp is the cokernel of $(UFn)\eta_C = \eta_{B\times C}n$. It follows that UFp is the direct image of p by $\eta_{B\times C}$.

Similarly, the morphism UFq is the direct image of q by $\eta_{B\times C}$. Since the pair (p,q) is a counion of regular epimorphisms, the pair (UFp,UFq) is a counion of regular epimorphisms, and this counion is disjoint since $\eta_{B\times C}$ is a semi-epimorphism. It is thus a product. The pair of regular epimorphisms, $(Fp : F(B\times C) \to FB,\ Fq : F(B\times C) \to FC)$, has as its counion, a regular epimorphism whose image under U is an isomorphism, thus it itself is an isomorphism since its kernel is null, its image under U being the null object. One shows similarly that the cointersection of (Fp, Fq) is null. It follows that the pair (Fp, Fq) is a product. The functor F thus preserves finite products and U is a morphism of quasi locally simple categories. ∎

3.7.2. Example. The fortgetful functor $\mathbb{R}$ngcBaLatNu $\to$ $\mathbb{R}$ngBaNu is a morphism of quasi locally simple categories.

3.7.3. Proposition. *A morphism of quasi locally simple categories preserves cokernels, semi-epimorphisms and objects of finite type.*

Proof : Let $U : \mathbb{A} \to \mathbb{B}$ be a morphism of quasi locally simple categories. It reflects the null morphisms since if $f : A \to B$ is a morphism of $\mathbb{A}$ such that $Uf = 0$, the kernel k of f is a normal monomorphism whose image under U is the kernel of Uf, but this latter is an isomorphism so k is an isomorphism (Proposition 3.7.1.), and $f \simeq 0$. Let us show that U preserves semi-epimorphisms. Let $f : A \to B$ be a semi-epimorphism of $\mathbb{A}$ and let $q : UB \to Q$ be the cokernel of Uf. If $n : N \to UB$ is the kernel of q, there exists a normal monomorphism $m : M \to B$ such that $Um \simeq n$. Since the functor U preserves cokernels of normal monomorphisms, the cokernel $k : B \to B/m$ of m has as its image under U, the cokernel q of $Um \simeq n$. It follows that $U(kf) = (Uk)(Uf) \simeq q(Uf) = 0$, thus $kf = 0$, $k = 0$ and $q = 0$. The morphism Uf is thus a semi-epimorphism. Let us show that U preserves cokernels. Let $f : A \to B$ be a morphism of $\mathbb{A}$ with cokernel $k : B \to K$. If $f = ng$ is the normal factorisation of f, then k is the cokernel of n. Since U preserves the cokernels of normal monomorphisms, Uk is the cokernel of Un. Since U preserves the semi-epimorphisms, Ug is a semi-epimorphism so Uk is the cokernel of $(Un)(Ug) = U(ng) = Uf$. The functor U preserves the objects of finite type, since if A is an object of finite type in $\mathbb{A}$, then the complete Heyting algebra $N_{\mathbb{A}}(A)$ is compact. The isomorphism $N_{\mathbb{A}}(A) \simeq N_{\mathbb{B}}(U(A))$ (3.7.4.) then proves that the Heyting algebra $N_{\mathbb{B}}(U(A))$ is compact, i.e. 1 is a compact element, and thus that the object $U(A)$ is of finite type (Propositions 3.5.8. and 3.6.3.). ∎

3.7.4. <u>The isomorphism</u> $\alpha : N_{\mathbb{A}} \to N_{\mathbb{B}}U$.

Let $U : \mathbb{A} \to \mathbb{B}$ be a morphism of quasi locally simple categories. We let $N_{\mathbb{A}}$ and $N_{\mathbb{B}}$ denote the functors, N, associated respectively to $\mathbb{A}$ and to $\mathbb{B}$ (Proposition 3.5.9.). For each object A of $\mathbb{A}$, one defines the mapping $\alpha_A : N_{\mathbb{A}}(A) \to N_{\mathbb{B}}(UA)$ by $\alpha_A(n) = Un$. It is an increasing mapping. Since the functor U uniquely lifts normal subobjects, the mappings α_A are isomorphisms of ordered sets, thus defining in fact a natural isomorphism $\alpha : N_{\mathbb{A}} \to N_{\mathbb{B}}U$.

3.8. <u>THE 2-EQUIVALENCE BETWEEN QUASI LOCALLY SIMPLE CATEGORIES AND LOCALLY SIMPLE CATEGORIES</u>.

Let $U : \mathbb{A} \to \mathbb{B}$ be a morphism of quasi locally simple categories. It induces a functor $U_{tf} : \mathbb{A}_{tf} \to \mathbb{B}_{tf}$ since it preserves objects of finite type and semi-epimorphisms (Proposition 3.7.3.). The functor $F : \mathbb{B} \to \mathbb{A}$, left adjoint to U, preserves cokernels, and thus semi-epimorphisms. It also preserves finitely presentable objects and thus their semi-quotients, that is to say, the objects of finite type (Proposition 3.6.2.). The functor F thus induces a functor $F_{tf} : \mathbb{B}_{tf} \to \mathbb{A}_{tf}$. Since the adjunction morphisms of (U,F) are semi-epimorphisms (Proposition 3.7.1.), the functor F_{tf} is left adjoint to U_{tf}. It follows that the functor $U_{tf} : \mathbb{A}_{tf} \to \mathbb{B}_{tf}$ is a morphism of locally simple categories (Definition 2.8.0.).

3.8.0. <u>Definition</u>. <u>A 2-morphism of quasi locally simple categories</u> is a natural transformation for which the source and the target are two morphisms of quasi locally simple categories and whose value on the coclassifier, Z, is a semi-epimorphism.

If $\alpha : U \to V$ is a 2-morphism of quasi locally simple categories with $U,V : \mathbb{A} \rightrightarrows \mathbb{B}$, then, for any object, A, of finite type in $\mathbb{A}$, the morphism $\alpha_A : UA \to VA$ is a semi-epimorphism, since, if we denote by $e_A : Z \to A$ the unique semi-epimorphism, the composite $\alpha_A(Ue_A) = (Ve_A)\alpha_Z$ is a semi-epimorphism. It follows that the 2-morphisms of quasi locally simple categories are stable by composition and constitute <u>the 2-category $\mathbb{Q}$uaLocSim, of quasi locally simple categories</u>. Any 2-morphism of quasi locally simple categories $\alpha : U \to V$ induces a natural transformation $\alpha_{tf} : U_{tf} \to V_{tf}$. If we denote by <u>$\mathbb{L}$ocSim the 2-category of locally simple categories</u>, morphisms of locally simple categories and natural transformations, we can define the <u>2-functor : Finite Type</u> :

$\mathbb{Q}$uaLocSim $\longrightarrow$ $\mathbb{L}$ocSim by assigning $\mathbb{A}_{tf}$ to $\mathbb{A}$, U_{tf} to U and α_{tf} to α.

3.8.1. Theorem. The 2-functor : Finite Type, establishes a 2-equivalence between the 2-category, $\mathbb{Q}$uaLocSim, of quasi locally simple categories, and the 2-category, $\mathbb{L}$ocSim, of locally simple categories.

Proof. Let $\mathbb{A}$ and $\mathbb{B}$ be two quasi locally simple categories. We denote by $\mathbb{H}om[\mathbb{A},\mathbb{B}]$ the category of morphisms and 2-morphisms from $\mathbb{A}$ to $\mathbb{B}$, $\mathbb{H}om[\mathbb{A}_{tf},\mathbb{B}_{tf}]$ that of morphisms and 2-morphisms from $\mathbb{A}_{tf}$ to $\mathbb{B}_{tf}$ and $(\ )_{tf} : \mathbb{H}om[\mathbb{A},\mathbb{B}] \to \mathbb{H}om[\mathbb{A}_{tf},\mathbb{B}_{tf}]$, the functor : "Finite Type". Denote by $J_{\mathbb{A}} : \mathbb{A}_{tf} \to \mathbb{A}$, the inclusion functor, $T : \mathbb{A} \to \mathbb{A}_{tf}$ its left adjoint and $\eta : 1_{\mathbb{A}} \to J_{\mathbb{A}}T$ the adjunction morphism. For each object A of $\mathbb{A}$, $\eta_A : A \to TA$ is a normal monomorphism whose cokernel is the morphism $q_A = T0 : TA \to T0 = Z$, since this property is easily checked for the objects of finite type (part (c) of the proof of Proposition 3.6.5.), and it can be extended by filtered colimits to all the objects. If $(\alpha,\beta) : U \rightrightarrows V$ is a pair of morphisms of $\mathbb{H}om[\mathbb{A},\mathbb{B}]$ such that $\alpha_{tf} = \beta_{tf}$, the relations $(V\eta_A)\beta_A = \beta_{TA}(U\eta_A) = \alpha_{TA}(U\eta_A) = (V\eta_A)\alpha_A$ imply that $\beta_A = \alpha_A$ and consequently that $\beta = \alpha$. The functor "Finite Type" is thus faithful. Let U and V be two objects of $\mathbb{H}om[\mathbb{A},\mathbb{B}]$ and $\gamma : U_{tf} \to V_{tf}$ a natural transformation. One has $(Vq_A)\gamma_{TA}(U\eta_A) = \gamma_Z(Uq_A)(U\eta_A) = \gamma_Z U(q_A\eta_A) = 0$. Since the functor V preserves kernels, $V\eta_A$ is the kernel of Vq_A and it follows that there is a unique morphism $\alpha_A : UA \to VA$ satisfying $(V\eta_A)\alpha_A = \gamma_{TA}(U\eta_A)$. It is immediate that this defines a natural transformation $\alpha : U \to V$ which extends $\gamma : U_{tf} \to V_{tf}$ and which is such that $\alpha_Z = \alpha_{T0} = \gamma_{T0}$ is a semi-epimorphism. It follows that $\alpha : U \to V$ is a morphism of $\mathbb{H}om[\mathbb{A},\mathbb{B}]$ such that $\alpha_{tf} = \gamma$. The functor "Finite Type" is thus full. Let $W : \mathbb{A}_{tf} \to \mathbb{B}_{tf}$ be an object of $\mathbb{H}om[\mathbb{A}_{tf},\mathbb{B}_{tf}]$. One defines the functor $V : \mathbb{A} \to \mathbb{B}$ by extending W by kernels : for an object A of $\mathbb{A}$, VA is the kernel of Wq_A and, for a morphism f of $\mathbb{A}$, Vf is the morphism induced by WTf. It is immediate that V is an extension of W, up to isomorphism, and that it preserves filtered colimits, regular epimorphisms and monomorphisms. Let us show that it reflects the null object. Let A be an object of $\mathbb{A}$ such that $VA = 0$. The image, Wq_A, of the regular epimorphism, q_A, under W, being a regular epimorphism whose kernel is null, is an isomorphism. Thus q_A is an isomorphism and $A = 0$. The functor V reflects null morphisms, since a morphism $f : A \to B$ such that $Vf = 0$

has a regular factorisation $f = mg$ with $g : A \to C$, a regular epimorphism, and $m : C \to B$, a monomorphism ; the image under V of this factorisation is the regular factorisation $0 = Vf = (Vm)(Vg)$, thus $Vg = 0$, $VC = 0$, $C = 0$ and $f = 0$. Let us show that the functor V has a left adjoint. Let (W,G,ν,ψ) the adjunction generated by the functor W. Let us show, first of all, that, for an object, B, of finite type of $\mathbb{B}$, the morphism $\nu_B : B \to WGB = VGB$ is universal from B to V. Let A be an object of $\mathbb{A}$ and $g : B \to VA$. There exists a unique morphism $h : GB \to TA$ satisfying $(Vh)\nu_B = (V\eta_A)g$. The relation $V(q_A h)\nu_B = (Vq_A)(Vh)\nu_B = (Vq_A)(V\eta_A)g = V(q_A\eta_A)g = 0$ together with the fact that ν_B is a semi-epimorphism, implie that $V(q_A h) = 0$ so $q_A h = 0$. Then there exists a unique morphism $f : GB \to A$ satisfying $\eta_A f = h$, thus satisfying $(V\eta_A)(Vf)\nu_B = V(\eta_A f)\nu_B = (Vh)\nu_B = (V\eta_A)g$ and consequently $(Vf)\nu_B = g$. As a result, the functor V has a partial left adjoint defined on the full subcategory of $\mathbb{B}$ having as its objects the objects of finite type, and this adjoint induces the functor G on $\mathbb{B}_{tf}$. This partial left adjoint extends, by filtered colimits, to a functor $\bar{G}$, left adjoint to the functor V and satisfying $\bar{G}J_{\mathbb{B}} = J_{\mathbb{A}}G$. Since the functors G and $J_{\mathbb{A}}$ preserve finite products so does the functor $\bar{G}J_{\mathbb{B}}$. Any finite product in $\mathbb{B}$, being a filtered colimit of finite products of finitely presentable objects, will thus be preserved by $\bar{G}$. The functor V is thus a morphism of quasi locally simple categories. This completes the proof that the "Finite Type" functor : $\mathbb{H}\mathrm{om}[\mathbb{A},\mathbb{B}] \to \mathbb{H}\mathrm{om}[\mathbb{A}_{tf}, \mathbb{B}_{tf}]$ is an equivalence of categories. Theorem 3.6.7. shows that the 2-functor "Finite Type" : $\mathbb{Q}$uaLocSim $\to$ $\mathbb{L}$ocSim is essentially surjective and consequently is a 2-equivalence of 2-categories. ∎

3.9. REPRESENTATION OF OBJECTS BY LOCALLY BOOLEAN SHEAVES OF SIMPLE OBJECTS.

3.9.0. Theorem. A quasi locally simple category is algebraic in the sense of Gabriel-Ulmer [16] and thus is equivalent to an I-algebraic category in the sense of Bénabou [6].

Proof : Let $\mathbb{A}$ be a quasi locally simple category. The category $\mathbb{A}_{tf}$ is locally simple (Theorem 3.6.6.) thus is algebraic in the sense of Gabriel-Ulmer (Theorem 2.4.0.). It is then easy to see that the category $\mathbb{A}^*_{tf}$ is algebraic, and consequently, that the category

$\mathbb{A}$, which is equivalent to $\mathbb{A}^*_{tf}$, is as well. ∎

A quasi locally simple category $\mathbb{A}$ is regular in the sense of Barr [2]. Hence, for any object A of $\mathbb{A}$, the lattice, N(A), of normal subobjects of A, the lattice E(A) of equivalence relations on A, and the dual of the lattice R(A) of regular quotient objects of A, are isomorphic, (Proposition 3.5.3.).

3.9.1. Examples. The category $\mathbb{R}$ngcBaNu does not look to be algebraic in the sense of Lawvere. It is, however, since one can define a non-unitary commutative Baer ring as being a commutative ring having a unitary operation e satisfying the four axioms :

(1) $e(0) = 0$

(2) $e(x^2) = e(x)$

(3) $xe(x) = x$

(4) $e(xy) = e(x)\, e(y)$

The category $\mathbb{R}$ngcBaNu, then, can be seen to be a variety in the category, $\mathbb{A}$n$\mathbb{E}$Nu.

Similarly $\mathbb{R}$ngcStLatProjNu is algebraic in the sense of Lawvere since one can define a non-unitary projectable strongly lattice-ordered ring as being a strongly lattice-ordered commutative ring having a unary operation e which satisfies the four axioms :

(1) $e(0) = 0$

(2) $e(|x|) = e(x)$

(3) $xe(x) = x$

(4) $e(|x| \wedge |y|) = e(x)\, e(y)$

The category $\mathbb{R}$ngcStLatProjNu can then be seen to be a variety in the category $\mathbb{R}$ngcEStLatNu.

A simple object of $\mathbb{A}$ is an object, S, having exactly two regular quotient objects, or, what amounts to the same, which has exactly two normal subobjects, i.e. so that $N(S) = \{0,1\} = 2$. A morphism between two simple objects is either null or a monomorphic semi-epimorphism, since, in the regular factorisation [2], $f = mq$, of such a non-null morphism f, the regular quotient q will be an isomorphism, so f will be a monomorphism, whilst in the normal factorisation $f = ns$, the subobject n is an isomorphism so f is a semi-epimorphism. The category $\mathbb{S}im\,\mathbb{A}$, of simple objects of $\mathbb{A}$ is the subcategory of $\mathbb{A}$ having, as its objects, the simple objects and, as its morphisms, the non-null morphisms. Following the notation of (2.6.0.), the full subcategory of simple objects in $\mathbb{A}_{tf}$ will be denoted $\mathbb{S}im\,\mathbb{A}_{tf}$.

3.9.2. Proposition. The categories $\mathbb{S}im\,\mathbb{A}$ and $\mathbb{S}im\,\mathbb{A}_{tf}$ coincide.

Proof : The null object is not simple as it has only one normal subobject. A simple object S of $\mathbb{A}$ is thus non-null and, consequently, there is a finitely presentable object A and a non-null morphism $f : A \to S$. In the normal factorisation $f = ng$, the morphism n is non-null thus is an isomorphism, and it follows that f is a semi-epimorphism, so that S is of finite type (Proposition 3.6.2.). The object S is simple in $\mathbb{A}_{tf}$ since the inclusion functor $J : \mathbb{A}_{tf} \to \mathbb{A}$ preserves regular quotient objects and reflects isomorphisms. Conversely, since $\mathbb{A}_{tf}$ is a subcategory of $\mathbb{A}$ closed under regular quotients, any simple object of $\mathbb{A}_{tf}$ is a simple object of $\mathbb{A}$. The simple objects of $\mathbb{A}$ are thus the same as those of $\mathbb{A}_{tf}$, and so, $\mathbb{S}im\,\mathbb{A} = \mathbb{S}im\,\mathbb{A}_{tf}$. ■

3.9.3. Notation.

By abuse of language, the category $\mathbb{S}hvLocBoolSim\,\mathbb{A}_{tf}$ of locally Boolean sheaves of simple objects of finite type of $\mathbb{A}$ will be called the category of locally Boolean sheaves of simple objects of $\mathbb{A}$, and will be denoted $\mathbb{S}hvLocBoolSim\,\mathbb{A}$. The category $\mathbb{S}hvEspLocBoolSim\,\mathbb{A}$ is defined similarly.

3.9.4. Theorem. Any quasi locally simple category $\mathbb{A}$ is equivalent to the category, $\mathbb{S}hvLocBoolSim\,\mathbb{A}$, of locally Boolean sheaves of simple objects of $\mathbb{A}$.

Proof : The category $\mathbb{S}hvLocBoolSim\,\mathbb{A}$ is quasi locally simple (Theorem 3.4.2.) and the category of its objects of finite type is the category $\mathbb{S}hvBoolSim\,\mathbb{A}$ (cf. 3.6.1. and 3.5.10.2.), which coincides with the category $\mathbb{S}hvBoolSim\,\mathbb{A}_{tf}$ (notation 3.9.3.). By Theorem 2.7.0., the category $\mathbb{S}hvBoolSim\,\mathbb{A}_{tf}$ is equivalent to the category $\mathbb{A}_{tf}$. By Theorem 3.8.1., the two categories $\mathbb{S}hvLocBoolSim\,\mathbb{A}$ and $\mathbb{A}$ are equivalent. ■

In order to get an equivalence between the category $\mathbb{A}$ and the category

$\mathcal{S}hvEspLocBool\mathcal{S}im\,\mathbb{A}$, one extends, by normal monomorphic filtered colimits, the structural sheaf functor $\Sigma_{tf} : \mathbb{A}_{tf} \to \mathcal{S}hvEspBoolSim\,\mathbb{A}_{tf}$ relative to the locally simple category $\mathbb{A}_{tf}$ (2.7.). This extented functor is the structural sheaf functor relative to $\mathbb{A}$ and is denoted $\Sigma(-) = (Spec_{Max}(-),(\tilde{-}))$ where $Spec_{Max} : \mathbb{A}^{op} \to \mathbb{E}spLocBool$ is the maximal spectrum functor relative to $\mathbb{A}$, and $(\tilde{-})$ is the structural sheaf on $Spec_{Max}(-)$. For each object A of $\mathbb{A}$, $Spec_{Max}(A) \simeq \varprojlim_{X \in N_{tf}(A)} Spec_{Max}(X)$ $\simeq \varprojlim_{X \in N_{tf}(A)} Spec_{Max}\, E_o(X)$ (Proposition 2.6.4.) $\simeq$ $Spec_{Max}(\varinjlim_{X \in N_{tf}(A)} E_o(X)) \simeq Spec_{Max}(\varinjlim_{X \in N_{tf}(A)} N_{tf}(X)) \simeq Spec_{Max}(N_{tf}(A))$. For a normal subobject of finite type $n : N \to A$ of A, one has $\tilde{A}(n) = A/n' \simeq N$ and, for a normal subobject $n_1 : N_1 \to A$ satisfying $n_1 k = n$ with $k : N \to N_1$, the restriction morphism, $\tilde{A}(n_1) \to \tilde{A}(n)$, is the unique retraction of k (Proposition 3.5.5.). From Proposition 3.5.8., one obtains :

3.9.5. Proposition. The maximal spectrum of an object A of $\mathbb{A}$ is a locally Boolean topological space whose points correspond to the maximal proper normal subobjects of A, whose set of open sets is a complete Heyting algebra isomorphic to the Heyting algebra $N(A)$ of normal subobjects of A, and whose set of open-compact subsets forms a non-unitary Boolean algebra isomorphic to the non-unitary Boolean algebra, $N_{tf}(A)$, of normal subobjects of finite type of A.

The structural sheaf $\tilde{A}$ on $Spec_{Max}(A)$ is defined on the open-compact subsets $D(N)$ by $\tilde{A}(D(N)) = N$.

By extending, by normal monomorphic filtered colimits, the global section functor $\Gamma_{tf} : \mathcal{S}hvBoolSim\,\mathbb{A}_{tf} \to \mathbb{A}_{tf}$ relative to the locally simple category $\mathbb{A}_{tf}$ (2.7), one gets a functor $\Gamma_o : \mathcal{S}hvLocBoolSim\,\mathbb{A} \to \mathbb{A}$ called the global sections with compact support functor, which is an equivalence of categories quasi-inverse to Σ. Each object (B,F) of $\mathcal{S}hvLocBoolSim\,\mathbb{A}$ is a normal monomorphic filtered colimit of objects of $\mathcal{S}hvBoolSim\,\mathbb{A}$ in the form $(B,F) = \varinjlim_{b \in B}((b),F_{(b)})$. For each $b \in B$, the image of the normal subobject , $(1,0) : ((b),F_{(b)}) \to ((b),F_{(b)}) \times ((b)',F_{(b)'}) = (B,F)$, under the functor Γ_o, is the normal subobject $(1,0) : Fb \to Fb \times (Fb)' = F1$, which is the subobject of global sections of F with support in b. The value of $\Gamma_o(B,F)$ is then $\varinjlim_{b \in B} \Gamma_o((b),F_{(b)}) = \varinjlim_{b \in B} Fb$, the normal filtered union of subobjects of global sections of F with support in b, where b runs through B, that is to say, the object of global sections with compact support of F.

3.10. QUASI LOCALLY SIMPLE MONADS.

3.10.0. Proposition. For a morphism, $U : \mathbb{A} \to \mathbb{B}$, of quasi locally simple categories, inducing the morphism $U_{tf} : \mathbb{A}_{tf} \to \mathbb{B}_{tf}$ of locally simple categories, one has the following equivalences :

(1) U is faithful $\Leftrightarrow$ U_{tf} is faithful.

(2) U lifts subobjects $\Leftrightarrow$ U_{tf} lifts subobjects.

(3) U reflects isomorphisms $\Leftrightarrow$ U_{tf} reflects isomorphisms $\Leftrightarrow$ U_{tf} is monadic $\Leftrightarrow$ U is monadic.

(4) U is fully faithful $\Leftrightarrow$ U_{tf} is fully faithful.

Proof : The notions considered in this proposition being definable in any 2-category, the results follow from the 2-equivalence given in Theorem 3.8.1. But let us give a direct proof. The adjunction $(U,F,\eta,\varepsilon) : \mathbb{A} \to \mathbb{B}$ induces the adjunction $(U_{tf}, F_{tf}, \eta_{tf}, \varepsilon_{tf}) : \mathbb{A}_{tf} \to \mathbb{B}_{tf}$ and the functor U preserves filtered colimits. Any object A of $\mathbb{A}$ is a filtered colimit of a diagram whose objects are in $\mathbb{A}_{tf}$, say $A = \varinjlim_{i\in\mathbb{I}} A_i$. Then $\varepsilon_A = \varinjlim_{i\in\mathbb{I}} \varepsilon_{A_i}$. Since the epimorphisms, regular epimorphisms and isomorphisms are stable under filtered colimits, it follows that ε is, respectively, an epimorphism, pointwise a regular epimorphism, or, an isomorphism if, and only if, ε_{tf} is. It follows that there are equivalences : U is faithful $\Leftrightarrow$ ε is an epimorphism $\Leftrightarrow$ ε_{tf} is an epimorphism $\Leftrightarrow$ U_{tf} is faithful. Similarly : U reflects isomorphisms $\Leftrightarrow$ ε is pointwise a regular epimorphism $\Leftrightarrow$ ε_{tf} is pointwise a regular epimorphism $\Leftrightarrow$ U_{tf} reflects isomorphisms. And also : U is fully faithful $\Leftrightarrow$ ε is an isomorphism $\Leftrightarrow$ ε_{tf} is an isomorphism $\Leftrightarrow$ U_{tf} is fully faithful. Analogously : U reflects subobjects $\Leftrightarrow$ η is pointwise a regular epimorphism $\Leftrightarrow$ η_{tf} is pointwise a regular epimorphism $\Leftrightarrow$ U_{tf} reflects subobjects. The two equivalences : U reflects isomorphisms $\Leftrightarrow$ U is monadic, and : U_{tf} reflects isomorphisms $\Leftrightarrow$ U_{tf} is monadic, follow form the monadicity criterium of Duskin [35, Proposition 21.5.13.], since the categories are algebraic in the sense of Gabriel-Ulmer and the functor U is regular [2]. ■

The monad $\mathbb{T} = (T,\eta,\mu)$ on $\mathbb{B}$ generated by a morphism of quasi locally simple categories, $U : \mathbb{A} \to \mathbb{B}$, is such that T preserves filtered colimits, cokernels (Proposition 3.7.3.) and finite products (Definition 3.7.0.), and the natural transformation η is a semi-epimorphism, i.e. the morphisms η_A are semi-epimorphisms (Proposition 3.7.1.).

3.10.1. Definition. A monad $\mathbb{T} = (T,\eta,\mu)$ on $\mathbb{A}$ is quasi locally simple if the category $\mathbb{A}$ is quasi locally simple, the functor T preserves filtered colimits, cokernels or finite products, and the natural transformation η is a semi-epimorphism.

3.10.2. Justification. Let us show that, for a monad $\mathbb{T} = (T,\eta,\mu)$ on a quasi locally simple category $\mathbb{A}$, such that T preserves filtered colimits and η is a semi-epimorphism, the two properties : T preserves cokernels, and : T preserves finite products, are equivalent. The functor T preserves semi-epimorphisms since if $f : B \to C$ is a semi-epimorphism, the relation $(Tf)\eta_B = \eta_C f$ with η_C a semi-epimorphism, implies that Tf is a semi-epimorphism. As the morphism $\eta_0 : 0 \to T0$ is a semi-epimorphism, one has $T0 \simeq 0$. Suppose that T preserves cokernels. Let $(p : B\times C \to B,\ q : B\times C \to C)$ be the product of B and C. Denote $n = (1,0) : B \to B\times C$ and $m = (0,1) : C \to B\times C$. As the morphism p is the cokernel of m, the morphism Tp is the cokernel of Tm and also cokernel of $(Tm)\eta_C = \eta_{B\times C}\, m$, since η_C is a semi-epimorphism. It follows that Tp is the direct image of p by $\eta_{B\times C}$. Similarly one shows that Tq is the direct image of q by $\eta_{B\times C}$, and, consequently, the pair (Tp,Tq) is a product since it is the direct image of a product by a semi-epimorphism. Conversely suppose that T preserves finite products. Then T preserves direct factors and their kernels. By passage to filtered colimits, T preserves all regular epimorphisms and their kernels. Let $f : B \to C$ be a morphism whose cokernel is $q : C \to Q$. If $f = ns$ is the normal factorisation of f, then q is the cokernel of n, thus Tq is the cokernel of Tn and also of $(Tn)(Ts) = T(ns) = Tf$, since Ts is a semi-epimorphism. The functor T thus preserves cokernels. ∎

3.10.3. Theorem. The forgetful functor associated to a quasi locally simple monad, is a morphism of quasi locally simple categories.

Proof : Let $\mathbb{T} = (T,\eta,\mu)$ be a quasi locally simple monad on $\mathbb{A}$. It induces a monad $\mathbb{T}_{tf} = (T_{tf},\eta_{tf},\mu_{tf})$ on the locally simple category $\mathbb{A}_{tf}$ such that the functor T_{tf} preserves filtered colimits and finite products. The forgetful functor $U^{\mathbb{T}_{tf}} : \mathbb{A}_{tf}^{\mathbb{T}_{tf}} \to \mathbb{A}_{tf}$ is a morphism of locally simple categories (Theorem 2.9.1.). It determines a morphism of quasi locally simple categories $V : \mathbb{X} \to \mathbb{A}$ (Theorem 3.8.1.) which is monadic (Proposition 3.10.0.) and which generates, on the category $\mathbb{A}$, a quasi locally simple monad $\mathbb{S} = (S,\varepsilon,\nu)$ inducing the monad $\mathbb{T}_{tf}$ on $\mathbb{A}_{tf}$ i.e. $\mathbb{S}_{tf} = \mathbb{T}_{tf}$. Then $\mathbb{S} = \mathbb{T}$ and $U^{\mathbb{T}} \simeq U^{\mathbb{S}} \sim V$ is a morphism of quasi locally simple categories. ∎

3.10.4. <u>Theorem</u>. <u>"Beck's Theorem". If $\mathbb{A}$ is a quasi locally simple category, a functor $U : \mathbb{X} \to \mathbb{A}$ is a monadic morphism of quasi locally simple categories if, and only if, the category $\mathbb{X}$ has a null object and generalised kernel pairs, and the functor U has a left adjoint, preserves filtered colimits and creates regular quotient objects.</u>

<u>Proof</u> : The conditions are necessary since a monadic morphism of quasi locally simple categories reflects isomorphisms, lifts normal subobjects (Proposition 3.7.1.), preserves cokernels (Proposition 3.7.3.) and thus lifts and creates regular quotient objects (2.9.0.1.). Conversely, suppose that U satisfies the announced conditions. U reflects isomorphisms and thus kernel pairs. It lifts quotients of equivalence relations since, for any equivalence relation $(m,n) : R \rightrightarrows X$ in $\mathbb{X}$, the coequaliser of (Um, Un) is of the form Uk where $k : X \to K$ is a regular epimorphism whose kernel pair is (m,n), since Uk has as its kernel pair (Um, Un), and it follows that k is the coequaliser of (m,n). Thus U creates quotients of equivalence relations. It follows then from Beck's theorem, as modified by Duskin [35, 21.5.13] that the functor U is monadic. The functor U, since it preserves filtered colimits, generates a monad of finite rank on the locally finitely presentable category $\mathbb{A}$ and, consequently, the category $\mathbb{X}$ is locally finitely presentable [16]. The functor U reflects the null object since, for any object X of $\mathbb{X}$ such that $UX = 0$, the regular epimorphism $X \to 0$ is an isomorphism since its image under U is. The functor U reflects null morphisms, since any morphism $f : X \to Y$ such that $Uf = 0$, has an isomorphism for its kernel, since its image under U has one, and so $f = 0$. Let $q : Y \to Q$ be the cokernel of a morphism $f : X \to Y$ of $\mathbb{X}$ and let $r : UY \to R$ be the cokernel of Uf. Then $Uq \leqslant r$. The regular epimorphism r is of the form Us with $s : Y \to S$, an epimorphisms of $\mathbb{X}$ satisfying $sf = 0$. Thus $s \leqslant q$ and $r = Us \leqslant Uq$ so $Uq \simeq r$. The functor U thus preserves cokernels. Consequently, the functor T preserves cokernels. The functor U lifts normal subobjects

since, for any object X of $\mathbb{X}$ and any normal monomorphism $f : A \to UX$, the cokernel $q : UX \to Q$ of f is of the form Uk, where $k : X \to K$ is a regular epimorphism whose kernel $n : N \to X$ is such that $Un \simeq f$. Let us denote by F the left adjoint to U and by $\eta : 1_{\mathbb{A}} \to UF$ the adjunction morphism. For each object A of $\mathbb{A}$, let $\eta_A = n_A s_A$, with $s_A : A \to \bar{A}$, $n_A : \bar{A} \to UFA$, be the normal factorisation of η_A. The morphism n_A is of the form $n_A = U(m_A)$ with $m_A : A_1 \to FA$ in $\mathbb{X}$. There is a morphism $t : FA \to A_1$ satisfying $(Ut)\eta_A = s_A$ and it is immediate that t is inverse to m_A, thus m_A is an isomorphism and $\eta_A \simeq s_A$ is a semi-epimorphism. It follows that the monad $\mathbb{T} = (T,\eta,\mu)$ is quasi locally simple (Definition 3.10.1.) and that $U \simeq U^{\mathbb{T}}$ is a monadic morphism of quasi locally simple categories (Theorem 3.10.3.). ■

3.10.5. Examples.

3.10.5.0. $\mathbb{R}$ngcRegStLatNu → $\mathbb{R}$ngcRegNu.

The forgetful functor U is algebraic. Let I be an ideal in a non-unitary strongly lattice-ordered regular ring A. The ideal, I, is generated by idempotents. If $x \in I$, there is an idempotent $e \in I$ satisfying $x = xe$. Let $y \in A$ be such that $|y| \leq |x|$. There is an idempotent f such that $y = yf$. Then $|y-ye| = |yf-yfe| = |y|\cdot|f-fe| \leq |x|\cdot|f-fe| = |xf-xfe| = 0$. Thus $y = ye \in I$. The ideal I is thus absolutely convex. Moreover if $x \in I$ and $y \in I$, there is an idempotent $g \in I$ satisfying $x = xg$ and $y = yg$ thus $x \vee y = gx \vee gy = g(x \vee y)$ and consequently $x \vee y \in I$. The ideal, I, is thus closed under finite joins. It follows that the quotient ring, A/I, is strongly lattice-ordered and that the canonical projection $A \to A/I$ is a regular quotient object in $\mathbb{R}$ngcRegStLatNu. The functor U thus lifts regular quotient objects. It is a monadic morphism of quasi locally simple categories, by Theorem 3.10.4.

Using "Beck's Theorem", one shows easily that the following forgetful functors are monadic morphisms of quasi locally simple categories :

3.10.5.1. $\mathbb{R}$ngcBaLatNu → $\mathbb{R}$ngcBaNu.

3.10.5.2. $\mathbb{R}$ngcEStLatNu → $\mathbb{R}$ngcENu

3.10.5.3. $\mathbb{R}$ngcDifRegNu → $\mathbb{R}$ngcRegNu

3.10.5.4. $\mathbb{A}$lgcRegNu(R) → $\mathbb{R}$ngcRegNu.

$\mathbb{A}$lgcRegNu(R) is the category of non-unitary regular commutative algebras over a non-unitary regular commutative ring, R. The forgetful functor U with values in $\mathbb{R}$ngcRegNu is algebraic in the sense of Lawvere. It is thus monadic and preserves filtered colimits. It is immediate that it creates regular quotient objects, since any ideal of a regular R-algebra is stable under multiplication by the elements of R. The category $\mathbb{R}$ngcRegNu(R) is thus quasi locally simple, and the functor U is a monadic morphism of quasi locally simple categories. Starting with any of the quasi locally simple categories of rings, one constructs, in this manner, quasi locally simple categories of algebras.

3.10.6. Notation. A quasi locally simple subcategory is a subcategory for which the inclusion functor is a morphism of quasi locally simple categories.

3.10.7. Proposition. If $\mathbb{A}$ is a quasi locally simple category, a full subcategory $\mathbb{X}$ of $\mathbb{A}$ is a quasi locally simple subcategory if, and only if, it is a reflexive subcategory of $\mathbb{A}$ which is closed under filtered colimits and, either regular quotient objects or normal subobjects.

Proof : As a reflexive subcategory is monadic, the result follows from Theorem 3.10.4. and the following equivalences : $\mathbb{X}$ is closed under regular quotient objects <=> $\mathbb{X}$ is closed under finitely generated regular quotients objects <=> $\mathbb{X}$ is closed under normal subobjects of finite type (cf. Proposition 3.5.5.) <=> $\mathbb{X}$ is closed under normal subobjects. ■

3.10.8. Example. The category $\mathbb{R}$ngcDifRegPerfNu(p) is a quasi locally simple subcategory of the category $\mathbb{R}$ngcDifRegNu.

3.11. QUASI LOCALLY SIMPLE QUOTIENT CATEGORIES.

Let $\mathbb{A}$ be a quasi locally simple category and $\mathbb{K}$ a full subcategory of $\mathbb{A}$ whose objects are finitely presentable.

3.11.0. Definition. The quotient category, $\mathbb{A}/\mathbb{K}$, of $\mathbb{A}$ by $\mathbb{K}$ is the full subcategory of $\mathbb{A}$ having as its objects, the objects A such that $\mathrm{Hom}_{\mathbb{A}}(K,A) = \{0\}$ for each object K of $\mathbb{K}$.

If $\mathbb{B}$ is a category with a null object 0, a functor $F : \mathbb{A} \to \mathbb{B}$ annihilates $\mathbb{K}$ if $FK = 0$ for every object K of $\mathbb{K}$.

3.11.1. Proposition. The quotient category $\mathbb{A}/\mathbb{K}$ is a full quasi locally simple subcategory of $\mathbb{A}$, whose reflector annihilates $\mathbb{K}$, and such that, any morphism of quasi locally simple categories, $U : \mathbb{B} \to \mathbb{A}$, whose left adjoint annihilates $\mathbb{K}$, can be factorised in a unique way through $\mathbb{A}/\mathbb{K}$ in a morphism of quasi locally simple categories from $\mathbb{B}$ to $\mathbb{A}/\mathbb{K}$.

Proof : The subcategory $\mathbb{A}/\mathbb{K}$ of $\mathbb{A}$ is closed under limits and subobjects. For each object A of $\mathbb{A}$, the regular quotient objects of A whose target is in $\mathbb{A}/\mathbb{K}$, form a solution set of morphisms from A to $\mathbb{A}/\mathbb{K}$ [33, p.117] and hence $\mathbb{A}/\mathbb{K}$ is a reflexive subcategory of $\mathbb{A}$ [33, Theorem 2, p.117]. As it is also closed under filtered colimits, it is a full quasi locally simple subcategory of $\mathbb{A}$ (Proposition 3.10.7.). Let $R : \mathbb{A} \to \mathbb{A}/\mathbb{K}$ be the reflector. For each object K of $\mathbb{K}$, the adjunction morphism $\eta_K : K \to RK$ is a semi-epimorphism (Proposition 3.7.1.) and null, thus $RK = 0$. For a morphism of quasi locally simple categories $U : \mathbb{B} \to \mathbb{A}$, whose left adjoint F annihilates $\mathbb{K}$, the objects UA satisfy $\mathrm{Hom}_{\mathbb{A}}(K,UA) \simeq \mathrm{Hom}_{\mathbb{B}}(FK,A) \simeq \mathrm{Hom}_{\mathbb{B}}(0,A) \simeq \{0\}$ for every object K of $\mathbb{K}$, hence $UA \in \mathbb{A}/\mathbb{K}$. It follows that U factorises uniquely in a morphism of quasi locally simple categories : $\mathbb{B} \to \mathbb{A}/\mathbb{K}$. ■

As the category $\mathbb{A}$ is equivalent to an I-algebraic category (Theorem 3.9.0.), the notion of a variety of $\mathbb{A}$ is naturally defined [9].

If $\mathbb{S}$ is a full subcategory of the category $\mathbb{S}\mathrm{im}\,\mathbb{A}$ of simple objects of $\mathbb{A}$, $\mathbb{L}\mathrm{oc}\,\mathbb{S}$ will denote the full subcategory of $\mathbb{A}$ whose objects are the objects A locally belonging to $\mathbb{S}$ i.e. such that the quotients of A by its maximal normal proper subobjects belong to $\mathbb{S}$ (2.10.0.).

3.11.2. Theorem. For a full subcategory, $\mathbb{X}$, of a quasi locally simple category, $\mathbb{A}$, the following assertions are equivalent :

(1) $\mathbb{X}$ is a quasi locally simple quotient category of $\mathbb{A}$.

(2) $\mathbb{X}$ is a variety of $\mathbb{A}$.

(3) $\mathbb{X}$ is a full quasi locally simple subcategory of $\mathbb{A}$, which is closed under subobjects.

(4) $\mathbb{X}$ is closed under products, subobjects, and regular quotient objects.

(5) <u>$\mathbb{X}$ is of the form $\mathbb{L}oc\ \mathbb{S}$ where $\mathbb{S}$ is a full subcategory of $\mathbb{S}im\ \mathbb{A}$ closed under ultraproducts and subobjects</u>.

<u>Proof</u> : We note that any finitely presentable object A of $\mathbb{A}$ is projective since, for a direct factor p, the mapping $Hom_{\mathbb{A}}(A,p)$ is surjective and that, for each regular epimorphism q, one has $q = \varinjlim_{i\in\mathbb{I}} q_i$ where $\mathbb{I}$ is filtered and the q_i are direct factors, thus, the mapping $Hom_{\mathbb{A}}(A,q) \simeq \varinjlim_{i\in\mathbb{I}} Hom_{\mathbb{A}}(A,q_i)$ is surjective.

<u>(1) <=> (2)</u>. The quotient category $\mathbb{A}/\mathbb{K}$ is the variety of $\mathbb{A}$ defined by the set of identities $\{(1_K,0) : K \rightrightarrows K\}_{K\in\mathbb{K}}$. Let $\mathbb{X}$ be the variety of $\mathbb{A}$ defined by the set of identities $\{(\omega_i,\mu_i) : X_i \rightrightarrows Y_i\}_{i\in I}$. For each $i \in I$, the coequaliser $q_i : Y_i \to Q_i$ of (ω_i,μ_i) is a finitely generated regular epimorphism, thus a direct factor forming, with its complement $p_i : Y_i \to P_i$, a product $(p_i : Y_i \to P_i,\ q_i : Y_i \to Q_i)$. Since Y_i is finitely presentable, P_i is finitely presentable. Let $\mathbb{K}$ be the full subcategory of $\mathbb{A}$ having as its objects, the various P_i for i in I. If A is an object of $\mathbb{X}$, any morphism $f : P_i \to A$ is null, since the morphism $fp_i : Y_i \to A$ necessarily factors through q_i and products are codisjoint in $\mathbb{A}$. Conversely, if A is an object such that any morphism $f : P_i \to A$ is null then any morphism $g : Y_i \to A$ factorises through q_i, since the pair $(\bar{p}_i,\bar{q}_i)$, direct image of (p_i,q_i) by g, is a counion of regular epimorphisms and $\bar{p}_i$ is necessarily null. It follows that $\mathbb{X} = \mathbb{A}/\mathbb{K}$.

<u>(2) <=> (3)</u>. By Proposition 3.11.1., it follows that (1) => (3) hence that (2) => (3). Let $\mathbb{X}$ be a full quasi locally simple subcategory of $\mathbb{A}$ closed under subobjects. The category $\mathbb{X}_{tf}$ is a full locally simple subcategory of $\mathbb{A}_{tf}$ (Proposition 3.10.0.) closed under subobjects. It is thus a variety of $\mathbb{A}_{tf}$ (Theorem 2.10.1.0.). Let $\{(\omega_i,\mu_i) : X_i \rightrightarrows Y_i\}_{i\in I}$ be a set of identities of $\mathbb{A}_{tf}$ defining $\mathbb{X}_{tf}$ and let $\mathbb{V}$ be the variety of $\mathbb{A}$ defined by the set of identities $\{(\omega_i,\mu_i)\}_{i\in I}$. One has $\mathbb{V}_{tf} \subset \mathbb{X}_{tf}$. If $A \in \mathbb{X}_{tf}$, any morphism $f : Y_i \to A$ factorises normally in the form $f = ns$ with s a semi-epimorphism and $n : N \to A$ a normal monomorphism of finite type, thus a split monomorphism (Proposition 3.5.5.), and consequently, $N \in \mathbb{X}_{tf}$, $s\omega_i = s\mu_i, f\omega_i = f\mu_i$ and so $A \in \mathbb{V}_{tf}$. As a result $\mathbb{X}_{tf} = \mathbb{V}_{tf}$. The two full quasi locally simple subcategories $\mathbb{X}$ and $\mathbb{V}$ are thus equal (Theorem 3.8.1.) and it follows that $\mathbb{X}$ is a variety of $\mathbb{A}$.

<u>(3) <=> (4)</u>. The implication (3) => (4) is immediate. If $\mathbb{X}$ is closed under products, subobjects and regular quotient objects, then the category $\mathbb{X}_{tf}$ is closed in $\mathbb{A}_{tf}$ under products, subobjects, and regular quotient objects, thus is a full locally simple subcategory of $\mathbb{A}_{tf}$ (Proposition 2.10.1.10.). There is a full quasi locally simple subcategory $\mathbb{Y}$ of $\mathbb{A}$ such that $\mathbb{Y}_{tf} = \mathbb{X}_{tf}$ (Theorem 3.8.1.). Then $\mathbb{Y} \subset \mathbb{X}$, since any object of $\mathbb{Y}$ is a normal subobject of an object of $\mathbb{Y}_{tf} = \mathbb{X}_{tf} \subset \mathbb{X}$ and $\mathbb{X}$ is closed under subobjects. Moreover $\mathbb{X} \subset \mathbb{Y}$, as any object of $\mathbb{X}$ is a monomorphic filtered colimit of objects of $\mathbb{X}_{tf} = \mathbb{Y}_{tf} \subset \mathbb{Y}$, and $\mathbb{Y}$ is closed under filtered colimits. Consequently, $\mathbb{X} = \mathbb{Y}$, and it follows that $\mathbb{X}$ is a full quasi locally simple subcategory of $\mathbb{A}$, which proves the implication (4) => (3).

<u>(3) <=> (5)</u>. If $\mathbb{X}$ is a full quasi locally simple subcategory of $\mathbb{A}$ closed under subobjects, the subcategory $\mathbb{S}\text{im}\,\mathbb{X}$ is closed in $\mathbb{S}\text{im}\,\mathbb{A}$ under ultraproducts and subobjects and one has $\mathbb{X} = \mathbb{L}\text{oc}\mathbb{S}\text{im}\,\mathbb{X}$. Conversely, let $\mathbb{S}$ be a full subcategory of $\mathbb{S}\text{im}\,\mathbb{A}$ closed under ultraproducts and subobjects. If $n : N \to A$ is a finitely generated normal monomorphism, any simple regular quotient object of N is a simple regular quotient object of A, (Proposition 3.5.5.). By passage to filtered colimits, this property can be extended to all the normal monomorphisms of $\mathbb{A}$. The category $\mathbb{L}\text{oc}\,\mathbb{S}$ is thus closed under normal subobjects. The category $\mathbb{L}\text{oc}_{tf}\mathbb{S}$ of objects of $\mathbb{L}\text{oc}\,\mathbb{S}$ of finite type is a full locally simple subcategory of $\mathbb{A}_{tf}$ (Theorem 2.10.1.0.). Thus there exists a full quasi locally simple subcategory $\mathbb{Y}$ of $\mathbb{A}$ such that $\mathbb{Y}_{tf} = \mathbb{L}\text{oc}_{tf}\,\mathbb{S}$ (Theorem 3.8.1.). Then $\mathbb{Y} \subset \mathbb{L}\text{oc}\,\mathbb{S}$, since any object of $\mathbb{Y}$ is a normal subobject of an object of $\mathbb{Y}_{tf} = \mathbb{L}\text{oc}_{tf}\,\mathbb{S} \subset \mathbb{L}\text{oc}\,\mathbb{S}$ and $\mathbb{L}\text{oc}\,\mathbb{S}$ is closed under normal subobjects. However $\mathbb{L}\text{oc}\,\mathbb{S} \subset \mathbb{Y}$, since any object of $\mathbb{L}\text{oc}\,\mathbb{S}$ is a normal monomorphic filtered colimit of objects of $\mathbb{L}\text{oc}_{tf}\,\mathbb{S} = \mathbb{Y}_{tf} \subset \mathbb{Y}$ and $\mathbb{Y}$ is closed under filtered colimits. Thus $\mathbb{L}\text{oc}\,\mathbb{S} = \mathbb{Y}$ is a quasi locally simple subcategory of $\mathbb{A}$. ∎

3.11.3. <u>Examples</u>. The categories $\mathbb{R}$ngcRegNu, $\mathbb{R}$ngcRegNu(p), p-$\mathbb{R}$ngcNu, p^k-$\mathbb{R}$ngcNu, $\mathbb{B}$oolNu are varieties of $\mathbb{R}$ngcStRegNu. The category $\mathbb{R}$ngcBaStLatNu is a variety of $\mathbb{R}$ngcBaLatNu. The categories $\mathbb{R}$ngc$\mathbb{E}$Nu, $\mathbb{R}$ngcBaNu are varieties of $\mathbb{R}$ng$\mathbb{E}$Nu. The category $\mathbb{R}$ngcDifRegNu(p) is a variety of $\mathbb{R}$ngcDifRegNu.

Let $\mathbb{T}$ be an algebraic theory in the sense of Lawvere having two distinguished constants 0 and 1 and $\mathbb{M}$ be a full subcategory of $\mathbb{A}$lg($\mathbb{T}$). The category $\mathbb{S}$hvLocBoolTsep $\mathbb{M}$ <u>of totally separated locally</u>

Boolean sheaves of algebras from $\mathbb{M}$ is the full subcategory of $\mathbb{S}hvLocBoolTsep\mathbb{A}lg(\mathbb{T})$ (Theorem 3.4.3.) whose objects are the pairs (X,F) such that the stalks of F are in $\mathbb{M}$.

3.11.4. Theorem. If $\mathbb{T}$ is an algebraic theory having two distinguished constants and if $\mathbb{M}$ is a full subcategory of $\mathbb{A}lg(\mathbb{T})$ closed under ultraproducts and subobjects, the category $\mathbb{S}hvLocBoolTsep\,\mathbb{M}$ of totally separated locally Boolean sheaves of algebras from $\mathbb{M}$ is quasi locally simple. The categories of this form are precisely the quotient categories of $\mathbb{S}hvLocBoolTsep\mathbb{A}lg(\mathbb{T})$.

Proof : The category $\mathbb{S}hvLocBoolTsep\mathbb{A}lg(\mathbb{T})$ is quasi locally simple (Theorem 3.4.3.) and the category of its simple objects is equivalent to the category $\mathbb{A}lgMono(\mathbb{T})$ having, as its objects, the non-null $\mathbb{T}$-algebras, and, as its morphisms, the injective homomorphisms. By Theorem 3.11.2., the quotient categories of $\mathbb{S}hvLocBoolTsep\mathbb{A}lg(\mathbb{T})$ are the full subcategories of the form $\mathbb{L}oc\,\mathbb{M}$ where $\mathbb{M}$ is a full subcategory of $\mathbb{A}lgMono(\mathbb{T})$ closed under ultraproducts and subobjects. These are the categories of the form $\mathbb{S}hvLocBoolTsep\,\mathbb{M}$. ■

INDEX

REFERENCES

[1] R.F. ARENS et I. KAPLANSKY : Topological representation of algebras. Trans. Amer. Math. Soc. 63, 1948, pp. 457-481.

[2] M. BARR : Exact categories and categories of sheaves. Lecture Notes in Mathematics 236, pp. 1-120, Springer-Verlag, Berlin/Heidelberg/New-York, 1971.

[3] M. BARR : Atomic Toposes. J. Pure. Appl. Algebra 17, 1980, pp. 1-24.

[4] M. BARR : Abstract Galois Theory. J. Pure. Appl. Algebra 19, 1980, p. 21-42.

[5] M. BARR : Abstract Galois Theory II. J. Pure. Appl. Algebra 25, 1982, p. 227-247.

[6] J. BENABOU : Structures algébriques dans les catégories. Thesis of the University of Paris, 1966.

[7] G.W. BRUMFIEL : Partially Ordered Rings and Semi-Algebraic Geometry, London Mathematical Society Lecture Notes Series 37, Cambridge Univ. Press, London/New-York, 1979.

[8] Y. DIERS : Type de densité d'une sous-catégorie pleine. Ann. Soc. Sc. Bruxelles. Tome 90-I, 1976, pp. 25-47.

[9] Y. DIERS : Variétés d'une catégorie. Annales de la Société Scientifique de Bruxelles Tome 90-II, 1976, pp. 159-172.

[10] Y. DIERS : Familles universelles de morphismes. Ann. Soc. Sc. Bruxelles, Tome 93-III, 1979, pp. 175-195.

[11] Y. DIERS : Catégories localement multiprésentables. Arch. Math. 34, 1980, pp. 344-356.

[12] Y. DIERS : Sur les familles monomorphiques régulières de morphismes. Cahiers de Topologie et Géométrie Différentielle, Vol. XXI-4, 1980, pp. 411-425.

[13] Y. DIERS : Quelques constructions de catégories localement multiprésentables. Ann. Sc. Math. Québec IV, n°2, 1980, pp. 79-101.

[14] Y. DIERS : Some spectra relative to functors. J. Pure. Appl. Algebra 22, 1981, 57-74.

[15] Y. DIERS : Une description axiomatique des catégories de faisceaux de structures algébriques sur les espaces topologiques booléens. Advances in Mathematics 47, 1983, pp. 258-299.

[16] P. GABRIEL et F. ULMER : Lokal präsentierbare Kategorien. Lecture Notes in Mathematics, 221, Springer-Verlag, Berlin/Heidelberg/New-York, 1971.

[17] P. GABRIEL et M. ZISMAN : Calculus of Fractions and Homotopy Theory. Springer-Verlag, Berlin/Heidelberg/New-York 1967.

[18] G. GRÄTZER : Lattice Theory. W.H. Freeman, San Francisco, 1971.

[19] J.W. GRAY : Fibred and cofibred categories. in Proceedings of the Conference on Categorical Algebra, La Jolla, 1965, pp. 21-83, Springer-Verlag, Berlin/Heidelberg/New-York, 1966.

[20] A. GROTHENDIECK : Revêtements étales et groupe fondamental. (S.G.A.1), Lecture Notes in Math 224, Springer-Verlag, Berlin, 1971.

[21] A. GROTHENDIECK, M. ARTIN, J.L. VERDIER : Théorie des Topos et Cohomologie Etale des schémas. Lecture Notes in Mathematics 269, Springer-Verlag, Berlin/Heidelberg/New-York, 1972.

[22] K.H. HOFMANN : Representations of algebras by continuous sections. Bull. Amer. Math. Soc. 78, 1972, 291-373.

[23] P.T. JOHNSTONE : Topos Theory. Academic Press, London/New-York/San Francisco, 1977.

[24] K. KEIMEL : The representation of lattice-ordered groups and rings by sections in sheaves. Lecture Notes in Mathematics 248, Springer-Verlag, Berlin/Heidelberg/New-York, 1971, pp. 1-98.

[25] J.F. KENNISON : Integral domain type representations in sheaves and other topoï. Math. Z. 151, 1976, pp. 35-56.

[26] J.F. KENNISON : Triples and compact sheaf representations. J. Pure. Appl. Algebra 20, 1981, pp. 13-38.

[27] A. KOCK : Formally real local rings and infinitesimal stability. in "Topos Theoretic Methods in Geometry", various publications series 30, Aarhus Universitet, 1979.

[28] E.R. KOLCHIN : Differential algebra and algebraic groups. Academic Press, London/New-York/San Francisco, 1973.

[29] J. LAMBEK : On the representation of modules by sheaves of factor modules. Canad. Math. Bull. 14, 1971, 359-368.

[30] F.W. LAWVERE : Functorial semantics of algebraic theories. Proc. Nat. Acad. Sci. U.S.A. 50, 1963, pp. 869-873.

[31] F.E.J. LINTON : Some aspects of equational categories. Proceedings of the Conference on Categorical Algebra, La Jolla, 1965, pp. 84-95, Springer-Verlag, Berlin/Heidelberg/New-York, 1966.

[32] N.H. MacCOY et D. MONTGOMERY : A representation of generalised Boolean rings. Duke Math. J. 3, 1937, pp. 455-459.

[33] S. MAC LANE : Categories for the Working Mathematician. Springer-Verlag, New-York/Heidelberg/Berlin, 1971.

[34] J.V. NEUMANN : Regular Rings. Proc. Nat. Acad. Sci. U.S.A. 22, 1936, pp. 707-713.

[35] H. SCHUBERT : Categories. Springer-Verlag, Berlin/Heidelberg/New-York, 1972.

[36] F. ULMER : Locally α-Presentable and Locally α-Generated Categories, Lecture Notes in Mathematics 195, Springer-Verlag, Berlin/Heidelberg/New-York, 1971, pp. 230-247.

[37] D.H. VAN OSDOL : Sheaves in Regular Categories . Lecture Notes in Mathematics 236, Springer-Verlag, Berlin/Heidelberg/New-York, 1971, pp. 223-239.

* *

*

Vol. 1034: J. Musielak, Orlicz Spaces and Modular Spaces. V, 222 pages. 1983.

Vol. 1035: The Mathematics and Physics of Disordered Media. Proceedings, 1983. Edited by B.D. Hughes and B.W. Ninham. VII, 432 pages. 1983.

Vol. 1036: Combinatorial Mathematics X. Proceedings, 1982. Edited by L.R.A. Casse. XI, 419 pages. 1983.

Vol. 1037: Non-linear Partial Differential Operators and Quantization Procedures. Proceedings, 1981. Edited by S.I. Andersson and H.-D. Doebner. VII, 334 pages. 1983.

Vol. 1038: F. Borceux, G. Van den Bossche, Algebra in a Localic Topos with Applications to Ring Theory. IX, 240 pages. 1983.

Vol. 1039: Analytic Functions, Błażejewko 1982. Proceedings. Edited by J. Ławrynowicz. X, 494 pages. 1983

Vol. 1040: A. Good, Local Analysis of Selberg's Trace Formula. III, 128 pages. 1983.

Vol. 1041: Lie Group Representations II. Proceedings 1982–1983. Edited by R. Herb, S. Kudla, R. Lipsman and J. Rosenberg. IX, 340 pages. 1984.

Vol. 1042: A. Gut, K. D. Schmidt, Amarts and Set Function Processes. III, 258 pages. 1983.

Vol. 1043: Linear and Complex Analysis Problem Book. Edited by V.P. Havin, S.V. Hruščëv and N.K. Nikol'skii. XVIII, 721 pages. 1984.

Vol. 1044: E. Gekeler, Discretization Methods for Stable Initial Value Problems. VIII, 201 pages. 1984.

Vol. 1045: Differential Geometry. Proceedings, 1982. Edited by A.M. Naveira. VIII, 194 pages. 1984.

Vol. 1046: Algebraic K-Theory, Number Theory, Geometry and Analysis. Proceedings, 1982. Edited by A. Bak. IX, 464 pages. 1984.

Vol. 1047: Fluid Dynamics. Seminar, 1982. Edited by H. Beirão da Veiga. VII, 193 pages. 1984.

Vol. 1048: Kinetic Theories and the Boltzmann Equation. Seminar, 1981. Edited by C. Cercignani. VII, 248 pages. 1984.

Vol. 1049: B. Iochum, Cônes autopolaires et algèbres de Jordan. VI, 247 pages. 1984.

Vol. 1050: A. Prestel, P. Roquette, Formally p-adic Fields. V, 167 pages. 1984.

Vol. 1051: Algebraic Topology, Aarhus 1982. Proceedings. Edited by I. Madsen and B. Oliver. X, 665 pages. 1984.

Vol. 1052: Number Theory, New York 1982. Seminar. Edited by D.V. Chudnovsky, G.V. Chudnovsky, H. Cohn and M.B. Nathanson. V, 309 pages. 1984.

Vol. 1053: P. Hilton, Nilpotente Gruppen und nilpotente Räume. V, 221 pages. 1984.

Vol. 1054: V. Thomée, Galerkin Finite Element Methods for Parabolic Problems. VII, 237 pages. 1984.

Vol. 1055: Quantum Probability and Applications to the Quantum Theory of Irreversible Processes. Proceedings, 1982. Edited by L. Accardi, A. Frigerio and V. Gorini. VI, 411 pages. 1984.

Vol. 1056: Algebraic Geometry. Bucharest 1982. Proceedings, 1982. Edited by L. Bădescu and D. Popescu. VII, 380 pages. 1984.

Vol. 1057: Bifurcation Theory and Applications. Seminar, 1983. Edited by L. Salvadori. VII, 233 pages. 1984.

Vol. 1058: B. Aulbach, Continuous and Discrete Dynamics near Manifolds of Equilibria. IX, 142 pages. 1984.

Vol. 1059: Séminaire de Probabilités XVIII, 1982/83. Proceedings. Edité par J. Azéma et M. Yor. IV, 518 pages. 1984.

Vol. 1060: Topology. Proceedings, 1982. Edited by L. D. Faddeev and A.A. Mal'cev. VI, 389 pages. 1984.

Vol. 1061: Séminaire de Théorie du Potentiel. Paris, No. 7. Proceedings. Directeurs: M. Brelot, G. Choquet et J. Deny. Rédacteurs: F. Hirsch et G. Mokobodzki. IV, 281 pages. 1984.

Vol. 1062: J. Jost, Harmonic Maps Between Surfaces. X, 133 pages. 1984.

Vol. 1063: Orienting Polymers. Proceedings, 1983. Edited by J.L. Ericksen. VII, 166 pages. 1984.

Vol. 1064: Probability Measures on Groups VII. Proceedings, 1983. Edited by H. Heyer. X, 588 pages. 1984.

Vol. 1065: A. Cuyt, Padé Approximants for Operators: Theory and Applications. IX, 138 pages. 1984.

Vol. 1066: Numerical Analysis. Proceedings, 1983. Edited by D.F. Griffiths. XI, 275 pages. 1984.

Vol. 1067: Yasuo Okuyama, Absolute Summability of Fourier Series and Orthogonal Series. VI, 118 pages. 1984.

Vol. 1068: Number Theory, Noordwijkerhout 1983. Proceedings. Edited by H. Jager. V, 296 pages. 1984.

Vol. 1069: M. Kreck, Bordism of Diffeomorphisms and Related Topics. III, 144 pages. 1984.

Vol. 1070: Interpolation Spaces and Allied Topics in Analysis. Proceedings, 1983. Edited by M. Cwikel and J. Peetre. III, 239 pages. 1984.

Vol. 1071: Padé Approximation and its Applications, Bad Honnef 1983. Prodeedings. Edited by H. Werner and H.J. Bünger. VI, 264 pages. 1984.

Vol. 1072: F. Rothe, Global Solutions of Reaction-Diffusion Systems. V, 216 pages. 1984.

Vol. 1073: Graph Theory, Singapore 1983. Proceedings. Edited by K. M. Koh and H. P. Yap. XIII, 335 pages. 1984.

Vol. 1074: E. W. Stredulinsky, Weighted Inequalities and Degenerate Elliptic Partial Differential Equations. III, 143 pages. 1984.

Vol. 1075: H. Majima, Asymptotic Analysis for Integrable Connections with Irregular Singular Points. IX, 159 pages. 1984.

Vol. 1076: Infinite-Dimensional Systems. Proceedings, 1983. Edited by F. Kappel and W. Schappacher. VII, 278 pages. 1984.

Vol. 1077: Lie Group Representations III. Proceedings, 1982–1983. Edited by R. Herb, R. Johnson, R. Lipsman, J. Rosenberg. XI, 454 pages. 1984.

Vol. 1078: A.J.E.M. Janssen, P. van der Steen, Integration Theory. V, 224 pages. 1984.

Vol. 1079: W. Ruppert. Compact Semitopological Semigroups: An Intrinsic Theory. V, 260 pages. 1984

Vol. 1080: Probability Theory on Vector Spaces III. Proceedings, 1983. Edited by D. Szynal and A. Weron. V, 373 pages. 1984.

Vol. 1081: D. Benson, Modular Representation Theory: New Trends and Methods. XI, 231 pages. 1984.

Vol. 1082: C.-G. Schmidt, Arithmetik Abelscher Varietäten mit komplexer Multiplikation. X, 96 Seiten. 1984.

Vol. 1083: D. Bump, Automorphic Forms on GL (3,IR). XI, 184 pages. 1984.

Vol. 1084: D. Kletzing, Structure and Representations of Q-Groups. VI, 290 pages. 1984.

Vol. 1085: G.K. Immink, Asymptotics of Analytic Difference Equations. V, 134 pages. 1984.

Vol. 1086: Sensitivity of Functionals with Applications to Engineering Sciences. Proceedings, 1983. Edited by V. Komkov. V, 130 pages. 1984

Vol. 1087: W. Narkiewicz, Uniform Distribution of Sequences of Integers in Residue Classes. VIII, 125 pages. 1984.

Vol. 1088: A.V. Kakosyan, L.B. Klebanov, J.A. Melamed, Characterization of Distributions by the Method of Intensively Monotone Operators. X, 175 pages. 1984.

Vol. 1089: Measure Theory, Oberwolfach 1983. Proceedings. Edited by D. Kölzow and D. Maharam-Stone. XIII, 327 pages. 1984.

Vol. 1090: Differential Geometry of Submanifolds. Proceedings, 1984. Edited by K. Kenmotsu. VI, 132 pages. 1984.

Vol. 1091: Multifunctions and Integrands. Proceedings, 1983. Edited by G. Salinetti. V, 234 pages. 1984.

Vol. 1092: Complete Intersections. Seminar, 1983. Edited by S. Greco and R. Strano. VII, 299 pages. 1984.

Vol. 1093: A. Prestel, Lectures on Formally Real Fields. XI, 125 pages. 1984.

Vol. 1094: Analyse Complexe. Proceedings, 1983. Edité par E. Amar, R. Gay et Nguyen Thanh Van. IX, 184 pages. 1984.

Vol. 1095: Stochastic Analysis and Applications. Proceedings, 1983. Edited by A. Truman and D. Williams. V, 199 pages. 1984.

Vol. 1096: Théorie du Potentiel. Proceedings, 1983. Edité par G. Mokobodzki et D. Pinchon. IX, 601 pages. 1984.

Vol. 1097: R.M. Dudley, H. Kunita, F. Ledrappier, École d'Éte de Probabilités de Saint-Flour XII – 1982. Edité par P.L. Hennequin. X, 396 pages. 1984.

Vol. 1098: Groups – Korea 1983. Proceedings. Edited by A.C. Kim and B.H. Neumann. VII, 183 pages. 1984.

Vol. 1099: C.M. Ringel, Tame Algebras and Integral Quadratic Forms. XIII, 376 pages. 1984.

Vol. 1100: V. Ivrii, Precise Spectral Asymptotics for Elliptic Operators Acting in Fiberings over Manifolds with Boundary. V, 237 pages. 1984.

Vol. 1101: V. Cossart, J. Giraud, U. Orbanz, Resolution of Surface Singularities. Seminar. VII, 132 pages. 1984.

Vol. 1102: A. Verona, Stratified Mappings – Structure and Triangulability. IX, 160 pages. 1984.

Vol. 1103: Models and Sets. Proceedings, Logic Colloquium, 1983, Part I. Edited by G.H. Müller and M.M. Richter. VIII, 484 pages. 1984.

Vol. 1104: Computation and Proof Theory. Proceedings, Logic Colloquium, 1983, Part II. Edited by M.M. Richter, E. Börger, W. Oberschelp, B. Schinzel and W. Thomas. VIII, 475 pages. 1984.

Vol. 1105: Rational Approximation and Interpolation. Proceedings, 1983. Edited by P.R. Graves-Morris, E.B. Saff and R.S. Varga. XII, 528 pages. 1984.

Vol. 1106: C.T. Chong, Techniques of Admissible Recursion Theory. IX, 214 pages. 1984.

Vol. 1107: Nonlinear Analysis and Optimization. Proceedings, 1982. Edited by C. Vinti. V, 224 pages. 1984.

Vol. 1108: Global Analysis – Studies and Applications I. Edited by Yu.G. Borisovich and Yu.E. Gliklikh. V, 301 pages. 1984.

Vol. 1109: Stochastic Aspects of Classical and Quantum Systems. Proceedings, 1983. Edited by S. Albeverio, P. Combe and M. Sirugue-Collin. IX, 227 pages. 1985.

Vol. 1110: R. Jajte, Strong Limit Theorems in Non-Commutative Probability. VI, 152 pages. 1985.

Vol. 1111: Arbeitstagung Bonn 1984. Proceedings. Edited by F. Hirzebruch, J. Schwermer and S. Suter. V, 481 pages. 1985.

Vol. 1112: Products of Conjugacy Classes in Groups. Edited by Z. Arad and M. Herzog. V, 244 pages. 1985.

Vol. 1113: P. Antosik, C. Swartz, Matrix Methods in Analysis. IV, 114 pages. 1985.

Vol. 1114: Zahlentheoretische Analysis. Seminar. Herausgegeben von E. Hlawka. V, 157 Seiten. 1985.

Vol. 1115: J. Moulin Ollagnier, Ergodic Theory and Statistical Mechanics. VI, 147 pages. 1985.

Vol. 1116: S. Stolz, Hochzusammenhängende Mannigfaltigkeiten und ihre Ränder. XXIII, 134 Seiten. 1985.

Vol. 1117: D.J. Aldous, J.A. Ibragimov, J. Jacod, Ecole d'Été de Probabilités de Saint-Flour XIII – 1983. Édité par P.L. Hennequin. IX, 409 pages. 1985.

Vol. 1118: Grossissements de filtrations: exemples et applications. Seminaire, 1982/83. Edité par Th. Jeulin et M. Yor. V, 315 pages. 1985.

Vol. 1119: Recent Mathematical Methods in Dynamic Programming. Proceedings, 1984. Edited by I. Capuzzo Dolcetta, W.H. Fleming and T. Zolezzi. VI, 202 pages. 1985.

Vol. 1120: K. Jarosz, Perturbations of Banach Algebras. V, 118 pages. 1985.

Vol. 1121: Singularities and Constructive Methods for Their Treatment. Proceedings, 1983. Edited by P. Grisvard, W. Wendland and J.R. Whiteman. IX, 346 pages. 1985.

Vol. 1122: Number Theory. Proceedings, 1984. Edited by K. Alladi. VII, 217 pages. 1985.

Vol. 1123: Séminaire de Probabilités XIX 1983/84. Proceedings. Edité par J. Azéma et M. Yor. IV, 504 pages. 1985.

Vol. 1124: Algebraic Geometry, Sitges (Barcelona) 1983. Proceedings. Edited by E. Casas-Alvero, G.E. Welters and S. Xambó-Descamps. XI, 416 pages. 1985.

Vol. 1125: Dynamical Systems and Bifurcations. Proceedings, 1984. Edited by B.L.J. Braaksma, H.W. Broer and F. Takens. V, 129 pages. 1985.

Vol. 1126: Algebraic and Geometric Topology. Proceedings, 1983. Edited by A. Ranicki, N. Levitt and F. Quinn. V, 523 pages. 1985.

Vol. 1127: Numerical Methods in Fluid Dynamics. Seminar. Edited by F. Brezzi, VII, 333 pages. 1985.

Vol. 1128: J. Elschner, Singular Ordinary Differential Operators and Pseudodifferential Equations. 200 pages. 1985.

Vol. 1129: Numerical Analysis, Lancaster 1984. Proceedings. Edited by P.R. Turner. XIV, 179 pages. 1985.

Vol. 1130: Methods in Mathematical Logic. Proceedings, 1983. Edited by C.A. Di Prisco. VII, 407 pages. 1985.

Vol. 1131: K. Sundaresan, S. Swaminathan, Geometry and Nonlinear Analysis in Banach Spaces. III, 116 pages. 1985.

Vol. 1132: Operator Algebras and their Connections with Topology and Ergodic Theory. Proceedings, 1983. Edited by H. Araki, C.C. Moore, Ş. Strătilă and C. Voiculescu. VI, 594 pages. 1985.

Vol. 1133: K.C. Kiwiel, Methods of Descent for Nondifferentiable Optimization. VI, 362 pages. 1985.

Vol. 1134: G.P. Galdi, S. Rionero, Weighted Energy Methods in Fluid Dynamics and Elasticity. VII, 126 pages. 1985.

Vol. 1135: Number Theory, New York 1983–84. Seminar. Edited by D.V. Chudnovsky, G.V. Chudnovsky, H. Cohn and M.B. Nathanson. V, 283 pages. 1985.

Vol. 1136: Quantum Probability and Applications II. Proceedings, 1984. Edited by L. Accardi and W. von Waldenfels. VI, 534 pages. 1985.

Vol. 1137: Xiao G., Surfaces fibrées en courbes de genre deux. IX, 103 pages. 1985.

Vol. 1138: A. Ocneanu, Actions of Discrete Amenable Groups on von Neumann Algebras. V, 115 pages. 1985.

Vol. 1139: Differential Geometric Methods in Mathematical Physics. Proceedings, 1983. Edited by H. D. Doebner and J. D. Hennig. VI, 337 pages. 1985.

Vol. 1140: S. Donkin, Rational Representations of Algebraic Groups. VII, 254 pages. 1985.

Vol. 1141: Recursion Theory Week. Proceedings, 1984. Edited by H.-D. Ebbinghaus, G.H. Müller and G.E. Sacks. IX, 418 pages. 1985.

Vol. 1142: Orders and their Applications. Proceedings, 1984. Edited by I. Reiner and K. W. Roggenkamp. X, 306 pages. 1985.

Vol. 1143: A. Krieg, Modular Forms on Half-Spaces of Quaternions. XIII, 203 pages. 1985.

Vol. 1144: Knot Theory and Manifolds. Proceedings, 1983. Edited by D. Rolfsen. V, 163 pages. 1985.